Transition Metal Catalyzed Cross-Coupling Reactions

Transition Metal Catalyzed Cross-Coupling Reactions

Editor

Ioannis D. Kostas

MDPI • Basel • Beijing • Wuhan • Barcelona • Belgrade • Manchester • Tokyo • Cluj • Tianjin

Editor
Ioannis D. Kostas
Institute of Chemical Biology
National Hellenic Research
Foundation
Athens
Greece

Editorial Office
MDPI
St. Alban-Anlage 66
4052 Basel, Switzerland

This is a reprint of articles from the Special Issue published online in the open access journal *Catalysts* (ISSN 2073-4344) (available at: www.mdpi.com/journal/catalysts/special_issues/cross_coupl_react).

For citation purposes, cite each article independently as indicated on the article page online and as indicated below:

LastName, A.A.; LastName, B.B.; LastName, C.C. Article Title. *Journal Name* **Year**, *Volume Number*, Page Range.

ISBN 978-3-0365-2577-8 (Hbk)
ISBN 978-3-0365-2576-1 (PDF)

© 2021 by the authors. Articles in this book are Open Access and distributed under the Creative Commons Attribution (CC BY) license, which allows users to download, copy and build upon published articles, as long as the author and publisher are properly credited, which ensures maximum dissemination and a wider impact of our publications.

The book as a whole is distributed by MDPI under the terms and conditions of the Creative Commons license CC BY-NC-ND.

Contents

About the Editor vii

Preface to "Transition Metal Catalyzed Cross-Coupling Reactions" ix

Ioannis D. Kostas
Editorial *Catalysts*: Special Issue on Transition Metal Catalyzed Cross-Coupling Reactions
Reprinted from: *Catalysts* **2021**, *11*, 473, doi:10.3390/catal11040473 1

Ioannis D. Kostas and Barry R. Steele
Thiosemicarbazone Complexes of Transition Metals as Catalysts for Cross-Coupling Reactions
Reprinted from: *Catalysts* **2020**, *10*, 1107, doi:10.3390/catal10101107 3

Janwa El-Maiss, Tharwat Mohy El Dine, Chung-Shin Lu, Iyad Karamé, Ali Kanj, Kyriaki Polychronopoulou and Janah Shaya
Recent Advances in Metal-Catalyzed Alkyl–Boron ($C(sp^3)$–$C(sp^2)$) Suzuki-Miyaura Cross-Couplings
Reprinted from: *Catalysts* **2020**, *10*, 296, doi:10.3390/catal10030296 43

Martin Vareka, Benedikt Dahms, Mario Lang, Minh Hao Hoang, Melanie Trobe, Hansjörg Weber, Maximilian M. Hielscher, Siegfried R. Waldvogel and Rolf Breinbauer
Synthesis of a Bcl9 Alpha-Helix Mimetic for Inhibition of PPIs by a Combination of Electrooxidative Phenol Coupling and Pd-Catalyzed Cross Coupling
Reprinted from: *Catalysts* **2020**, *10*, 340, doi:10.3390/catal10030340 69

Miloslav Semler, Filip Horký and Petr Štěpnička
Synthesis of Alkynyl Ketones by Sonogashira Cross-Coupling of Acyl Chlorides with Terminal Alkynes Mediated by Palladium Catalysts Deposited over Donor-Functionalized Silica Gel
Reprinted from: *Catalysts* **2020**, *10*, 1186, doi:10.3390/catal10101186 79

About the Editor

Ioannis D. Kostas

Ioannis D. Kostas has a Degree in Chemistry (Uni. Thessaloniki, 1986) and a PhD in Organometallic Chemistry with C.G Screttas (Uni. Athens, 1991), and worked as a post-doc at the Vrije Universiteit/Amsterdam (1994–1995) with F. Bickelhaupt and at the Max-Planck-Institut für Kohlenforschung/Mülheim a.d. Ruhr (1995–1996) with M.T. Reetz. In 1996, he jointed the National Hellenic Research Foundation in Athens as a Researcher, in which he has been the "Research Director" since 2007. Since 2015, he has also been a Visiting Professor at the University of Thessaly. His research interests are focused on transition metal homogeneous catalysis and also the organic synthesis of bioactive compounds. He is pioneer in the use of thiosemicarbazones as ligands in Pd-catalyzed coupling reactions and also metalloporphyrins as catalysts in the Suzuki reaction and the hydrogenation of unsaturated aldehydes. He has been a coordinator or participant in 21 EU and National competitive programs.

Preface to "Transition Metal Catalyzed Cross-Coupling Reactions"

Transition metal-catalyzed cross-coupling reactions, such as Suzuki–Miyaura, Mizoroki–Heck, Negishi, Sonogashira, Kumada–Tamao–Corriu, Migita–Kosugi–Stille, Tsuji–Trost, and Buchwald–Hartwig reactions, have proved to be powerful tools for carbon–carbon as well as carbon–heteroatom bond formation in the development of synthetic methodologies for applications ranging from pharmaceuticals to materials. Intensive research efforts continue to be made to find ways of improving and expanding the scope of the processes, and the development of more efficient catalytic systems is a hot research topic of enormous academic and industrial interest.

This book, consisting of an editorial, two reviews and two articles, focuses on recent promising research and novel trends in the field of cross-coupling reactions, employing a range of different catalysts. The review by Kostas and Steele provides a survey of the research in the area of cross-coupling catalytic reactions with transition metal complexes based on the thiosemicarbazone unit and a discussion of the prospects for future developments. The other review by Polychronopoulou, Shaya and co-authors describes the progress made over the 21st century concerning the utilization of $C(sp^3)$–organoboranes as partners in metal-catalyzed $C(sp^3)$–$C(sp^2)$ cross couplings, such as B-alkyl Suzuki–Miyaura reactions. The article by Waldvogel, Breinbauer and co-authors demonstrates for the first time the synthetic potential of combining the electro-oxidative dehydrogenative cross-coupling of ortho-substituted phenols with Pd-catalyzed cross-coupling reactions. In the second article, Štěpnička and co-workers describe the preparation of palladium catalysts deposited over silica gel-bearing composite amide-donor functional moieties on the surface, which were evaluated in the Sonogashira-type cross-coupling of acyl chlorides with terminal alkynes.

In summary, this collection of publications represents some of the progress and recent trends in the expanding field of transition metal-catalyzed cross-coupling reactions. Special thanks are, of course, given to all the contributing authors and to the reviewers for their assessments and recommendations concerning the submitted manuscripts. I would also like to thank my colleague Dr. Barry R. Steele and the editorial team of *Catalysts* for their kind support and fast response.

Ioannis D. Kostas
Editor

Editorial

Editorial Catalysts: Special Issue on Transition Metal Catalyzed Cross-Coupling Reactions

Ioannis D. Kostas

Laboratory of Organic/Organometallic Chemistry & Catalysis, Institute of Chemical Biology, National Hellenic Research Foundation, Vas. Constantinou 48, 11635 Athens, Greece; ikostas@eie.gr

Citation: Kostas, I.D. Editorial *Catalysts*: Special Issue on Transition Metal Catalyzed Cross-Coupling Reactions. *Catalysts* **2021**, *11*, 473. https://doi.org/10.3390/catal11040473

Received: 10 March 2021
Accepted: 10 March 2021
Published: 7 April 2021

Publisher's Note: MDPI stays neutral with regard to jurisdictional claims in published maps and institutional affiliations.

Copyright: © 2021 by the author. Licensee MDPI, Basel, Switzerland. This article is an open access article distributed under the terms and conditions of the Creative Commons Attribution (CC BY) license (https://creativecommons.org/licenses/by/4.0/).

Transition metal catalyzed cross-coupling reactions have proved to be powerful tools for carbon–carbon as well as carbon–heteroatom bond formation in the development of synthetic methodologies for applications ranging from pharmaceuticals to materials. Intensive research efforts continue to be made into finding ways of improving and expanding the scope of the processes, and the development of more efficient catalytic systems is a hot research topic of enormous academic and industrial interest. Improvements in catalyst design are continually being made and have led to the use of milder conditions, immobilisation on solid supports, biphasic systems for ease of separation, more benign solvents, etc. Research in this area has led to a wide variety of very efficient and useful procedures which are now most often known by the names of the scientists who pioneered their use, such as Suzuki–Miyaura, Mizoroki–Heck, Negishi, Sonogashira, Kumada-Tamao-Corriu, Migita–Kosugi–Stille, Tsuji–Trost, Buchwald–Hartwig [1–8]. These procedures are mainly based on palladium although other metals have been shown to be effective in a number of cases.

This Special Issue, consisting of two reviews and two articles, focuses on recent promising research and novel trends in the field of cross-coupling reactions employing a range of different catalysts.

A review by Kostas and Steele provides a survey of the research in the area of cross-coupling catalytic reactions with transition metal complexes based on the thiosemicarbazone unit, and a discussion of the prospects for future developments [9]. Phosphanes have traditionally been the ligands of choice for transition metal catalysis but, since they can often be water- and air-sensitive, a number of efforts have been made to develop water and air-tolerant phosphane-free ligands. Thiosemicarbazone ligands possessing a wide variety of coordination modes via N, S or additional donors are excellent candidates for catalysis under phosphane-free conditions, and their use in coupling reactions was first reported in 2004 and 2005 for the Heck and Suzuki reactions, respectively [10,11]. The fact that the ligands are relatively readily accessible and that the complexes formed show good stability make them popular subjects for investigation. This review covers a large number of thiosemicarbazone-based catalysts for a variety of cross-coupling reactions, indicating the importance of these systems in catalysis.

Another review by Polychronopoulou, Shaya and co-authors describes progress during the 21st century concerning the utilization of $C(sp^3)$-organoboranes as partners in metal-catalyzed $C(sp^3)$–$C(sp^2)$ cross-couplings, such as B–alkyl Suzuki–Miyaura reactions [12]. Important topics of this review include the use of organic halides or pseudohalides as coupling partners, the strong interest in C–O-alkyl electrophiles, and progress in the syntheses of stable and isolable sp^3-boron reagents impacting the development of $C(sp^3)$–$C(sp^2)$ cross-couplings.

The article by Waldvogel, Breinbauer and co-authors demonstrates for the first time the synthetic potential of combining the electrooxidative dehydrogenative cross-coupling of ortho-substituted phenols with Pd-catalyzed cross-coupling reactions [13]. This synthetic methodology resulted Bcl9 quateraryl α-helix mimetics for inhibition of protein-protein interactions (PPIs), and it is expected that it will find applications in the synthesis of oligoarene structures. In the first step of the process, two phenols undergo electrooxidative dehydrogenative cross-coupling for the formation of 4,4′-biphenols. For the next step, the researchers found it necessary to activate the phenols as nonaflates that could be conveniently subjected to Pd-catalyzed cross-coupling reactions, whereas triflates show considerable issues in the subsequent Pd-catalyzed reactions due to their hydrolytic lability against bases. The nonaflate moiety serves as a leaving group for iterative Pd-catalyzed Suzuki-cross-coupling reactions with substituted pyridine boronic acids.

In a second article, Štěpnička and co-workers describe the preparation of palladium catalysts deposited over silica gel bearing composite amide-donor functional moieties on the surface [14]. These heterogeneous catalysts were evaluated in the Sonogashira-type cross-coupling of acyl chlorides with terminal alkynes producing synthetically useful 1,3-disubstituted prop-2-yn-1-ones. In general, they showed a good catalytic activity under relatively mild reaction conditions even without addition of a copper co-catalyst, but a careful optimization was required as the catalytic properties are significantly affected by the reaction conditions (solvent and base) and depend on the nature of the functional pendant on the support's surface.

In summary, this collection of publications represents some of the progress and recent trends in the expanding field of transition metal catalyzed cross-coupling reactions. I wish to thank the authors of the publications for their valuable contributions, my colleague Dr. Barry R. Steele, and the editorial team of Catalysts for their kind support and fast response.

Funding: This research received no external funding.

Conflicts of Interest: The author declares no conflict of interest.

References

1. Biffis, A.; Centomo, P.; Del Zotto, A.; Zecca, M. Pd Metal Catalysts for cross-couplings and related reactions in the 21st century: A critical review. *Chem. Rev.* **2018**, *118*, 2249–2295. [CrossRef] [PubMed]
2. Beletskaya, I.P.; Alonso, F.; Tyurin, V. The Suzuki-Miyaura reaction after the Nobel prize. *Coord. Chem. Rev.* **2019**, *385*, 137–173. [CrossRef]
3. Beletskaya, I.P.; Averin, A.D. New trends in the cross-coupling and other catalytic reactions. *Pure Appl. Chem.* **2017**, *89*, 1413–1428. [CrossRef]
4. Kanwal, I.; Mujahid, A.; Rasool, N.; Rizwan, K.; Malik, A.; Ahmad, G.; Shah, S.A.A.; Rashid, U.; Nasir, N.M. Palladium and copper catalyzed Sonogashira cross coupling an excellent methodology for C-C bond formation over 17 years: A review. *Catalysts* **2020**, *10*, 443. [CrossRef]
5. Choi, J.; Fu, G.C. Transition metal–catalyzed alkyl-alkyl bond formation: Another dimension in cross-coupling chemistry. *Science* **2017**, *356*, eaaf7230. [CrossRef] [PubMed]
6. Kumar, S. Recent advances in the Schiff bases and N-heterocyclic carbenes as ligands in the cross-coupling reactions: A comprehensive review. *J. Heterocycl. Chem.* **2019**, *56*, 1168–1230. [CrossRef]
7. Heravi, M.M.; Zadsirjan, V.; Hajiabbasi, P.; Hamidi, H. Advances in Kumada–Tamao–Corriu cross-coupling reaction: An update. *Monatshefte Chem. Chem. Mon.* **2019**, *150*, 535–591. [CrossRef]
8. Sain, S.; Jain, S.; Srivastava, M.; Vishwakarma, R.; Dwivedi, J. Application of palladium-catalyzed cross-coupling reactions in organic synthesis. *Curr. Org. Synth.* **2020**, *16*, 1105–1142. [CrossRef] [PubMed]
9. Kostas, I.D.; Steele, B.R. Thiosemicarbazone complexes of transition metals as catalysts for cross-coupling reactions. *Catalysts* **2020**, *10*, 1107. [CrossRef]
10. Kovala-Demertzi, D.; Yadav, P.N.; Demertzis, M.A.; Jasinski, J.P.; Andreadaki, F.J.; Kostas, I.D. First use of a palladium complex with a thiosemicarbazone ligand as catalyst precursor for the Heck reaction. *Tetrahedron Lett.* **2004**, *45*, 2923–2926. [CrossRef]
11. Kostas, I.D.; Andreadaki, F.J.; Kovala-Demertzi, D.; Prentjas, C.; Demertzis, M.A. Suzuki–Miyaura cross-coupling reaction of aryl bromides and chlorides with phenylboronic acid under aerobic conditions catalyzed by palladium complexes with thiosemicarbazone ligands. *Tetrahedron Lett.* **2005**, *46*, 1967–1970. [CrossRef]
12. El-Maiss, J.; El Dine, T.M.; Lu, C.-S.; Karamé, I.; Kanj, A.; Polychronopoulou, K.; Shaya, J. Recent advances in metal-catalyzed alkyl–boron (C(sp^3))–C(sp^2)) Suzuki-Miyaura cross-couplings. *Catalysts* **2020**, *10*, 296. [CrossRef]
13. Vareka, M.; Dahms, B.; Lang, M.; Hoang, M.H.; Trobe, M.; Weber, H.; Hielscher, M.M.; Waldvogel, S.R.; Breinbauer, R. Synthesis of a Bcl9 alpha-helix mimetic for inhibition of PPIs by a combination of electrooxidative phenol coupling and Pd-catalyzed cross coupling. *Catalysts* **2020**, *10*, 340. [CrossRef]
14. Semler, M.; Horký, F.; Štěpnička, P. Synthesis of alkynyl ketones by Sonogashira cross-coupling of acyl chlorides with terminal alkynes mediated by palladium catalysts deposited over donor-functionalized silica gel. *Catalysts* **2020**, *10*, 1186. [CrossRef]

Review

Thiosemicarbazone Complexes of Transition Metals as Catalysts for Cross-Coupling Reactions

Ioannis D. Kostas *[ID] and Barry R. Steele *

Laboratory of Organic/Organometallic Chemistry & Catalysis, Institute of Chemical Biology, National Hellenic Research Foundation, Vas. Constantinou 48, 11635 Athens, Greece
* Correspondence: ikostas@eie.gr (I.D.K.); bsteele@eie.gr (B.R.S.); Tel.: +30-2107273878 (I.D.K.); +30-2107273874 (B.R.S.)

Received: 2 September 2020; Accepted: 22 September 2020; Published: 24 September 2020

Abstract: Catalysis of cross-coupling reactions under phosphane-free conditions represents an important ongoing challenge. Although transition metal complexes based on the thiosemicarbazone unit have been known for a very long time, their use in homogeneous catalysis has been studied only relatively recently. In particular, reports of cross-coupling catalytic reactions with such complexes have appeared only in the last 15 years. This review provides a survey of the research in this area and a discussion of the prospects for future developments.

Keywords: thiosemicarbazone; metal complex; transition metal catalysis; cross-coupling reaction; Heck reaction; Suzuki reaction; Sonogashira reaction; Kumada reaction; Buchwald–Hartwig reaction

1. Introduction

Catalysis by means of transition metal complexes is now a well-established tool for the organic chemist, and the continued interest in the field has led to increasingly more effective and efficient systems for carrying out a wide range of reactions, both on a laboratory and on an industrial scale. The benefits of transition metal catalysis are that the reactions are often very clean and have very high turnovers, meaning that waste products are kept to a minimum, which is one of the precepts of Green Chemistry. In addition, improvements in catalyst design are continually being made and thus allow the use of milder conditions, immobilisation on solid supports, biphasic systems for ease of separation, more benign solvents, etc. Research in the area of transition metal catalysed carbon–carbon and carbon–heteroatom coupling reactions has led to a wide variety of very efficient and useful procedures which are now most often known by the names of the scientists who pioneered their use such as Suzuki–Miyaura, Mizoroki–Heck, Negishi, Sonogashira, Kumada–Tamao–Corriu, Migita–Kosugi–Stille, Tsuji–Trost, Buchwald–Hartwig [1–9]. These procedures are mainly based on palladium although other metals have been shown to be effective in a number of cases. Phosphane ligands have traditionally been the ligands of choice for transition metal catalysis and particularly so for coupling reactions. Such systems are generally rather stable and have been refined to a very great extent. However, since phosphanes can often be water- and air-sensitive, a number of efforts have been made to develop catalysts which avoid them and instead to employ ligands with C, N, O, or S donor groups [10], e.g., *N*-heterocyclic carbenes, carbocyclic carbenes, oxazolines, Schiff bases, amines, imidazoles, hydrazones, semicarbazones, thiosemicarbazones, thioureas, amidates, and so on. This review focuses on complexes of thiosemicarbazones and on how they can play a role in these developments.

The use of thiosemicarbazones, as well as other closely related chalcogen compounds, as ligands in metal complexes has proved to be a fruitful field of study for many years but initial reports on their application in catalysis did not appear until the 1990's [11–14], while their use in coupling reactions was only first reported several years later [15,16]. One of the primary motivations for research into these complexes has been the various areas in which they have been proposed for application. For example, apart from their activity as catalysts, which will be covered in more detail in this review, many thiosemicarbazone metal complexes have been widely studied as potential treatments for various types of cancer, for viral, bacterial, or fungal infections, and for neurodegenerative diseases, or for malaria [17–26]. Thiosemicarbazone metal complexes have also found potential application in medical imaging [27,28], while thiosemicarbazones themselves show promise as metal ion sensors and for the scavenging of metals due to their selective and specific coordination properties [28–34]. Although related compounds such as isothiosemicarbazones, dithiocarbazates, and selenosemicarbazones have so far found less application in the catalysis of coupling reactions and will not feature in this review, it is appropriate to mention that a number of metal complexes have nevertheless been reported and that they too have been studied in areas such as oxidation processes [35–37], cytotoxicity [38–47], antimicrobials [48], imaging [49,50], and antioxidants [43,51,52].

Another important aspect of thiosemicarbazones as ligands is the wide variety of coordination modes which can be adopted. Numerous structural studies have been carried out and this area has been the subject of a number of reviews [17,53–55]. An equilibrium mixture of thione (**I**) and thiol (**II**) tautomers exists in solution (Scheme 1). The simplest thiosemicarbazones, without any additional potential donor sites, can adopt a bidentate configuration either in their neutral form or in their deprotonated form as an anionic ligand. Many thiosemicarbazones, however, also have additional functionality which provides further potential donor sites and thus enables tridentate or higher degrees of denticity (Scheme 2). This becomes important when dealing with carbon–carbon coupling reactions, and especially those involving palladium complexes, since pincer-type Pd(II) complexes are known to efficiently catalyse carbon–carbon coupling reactions [56], and it has been hypothesised that the tridentate coordination of the pincer ligand stabilises the metal–carbon bond during the catalytic cycle [57].

Scheme 1. Tautomerism of thiosemicarbazones.

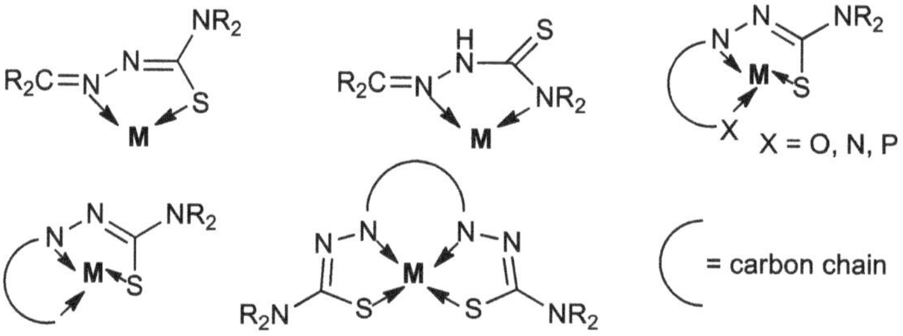

Scheme 2. Representative coordination modes of thiosemicarbazones.

This versatility of coordination, together with the relative ease with which these ligands can often be prepared, has provided a considerable impetus into the study of their metal complexes, and particularly how it can be exploited for the development of new catalysts. Reactions which have been studied using these systems include oxidation [58–74], transfer hydrogenation [75–77], reduction [14], silane alcoholysis [12,13,78], condensation reactions [79–81], and the cyclo propanation of olefins [82,83], as well as coupling reactions which are described here. Some of these have also previously been covered by Kumar et al. in reviews of the role of organochalcogen ligands in Mizoroki–Heck and Suzuki–Miyaura reactions [84–86]. The present review is organised according to the nature of the bond formed, i.e. carbon–carbon or carbon–heteroatom, with further subdivisions within these categories. It should be noted that, as with many other catalytic systems, the complexes used should usually more accurately be referred to as pre-catalysts since the active species are often formed in the reaction system. Indeed, there may be more than one active species formed, giving rise to a catalytic "cocktail" [87,88]. There have been numerous studies concerning the mechanism of coupling reactions catalysed by transition metal complexes which discuss the nature and formation of these cocktails as well as other related features of these reactions such as the aggregation of complexes or their de-aggregation, leaching effects, the role of nanoparticles and so on [89–92] but relatively little such work has been done, however, on thiosemicarbazone complexes. This aspect is therefore very much in its infancy and the present review will attempt to highlight the most significant studies in this area.

2. Carbon–Carbon Coupling Reactions

2.1. Mizoroki–Heck Reaction

The palladium catalysed coupling of alkenes and aryl halides was discovered independently by Mizoroki and Heck (Scheme 3). The numerous modifications and variations of this reaction have been so extensively reviewed that there is even a review of reviews on the subject [93]. Metal complexes of thiosemicarbazones that have been used as catalysts for this reaction are shown in Figure 1.

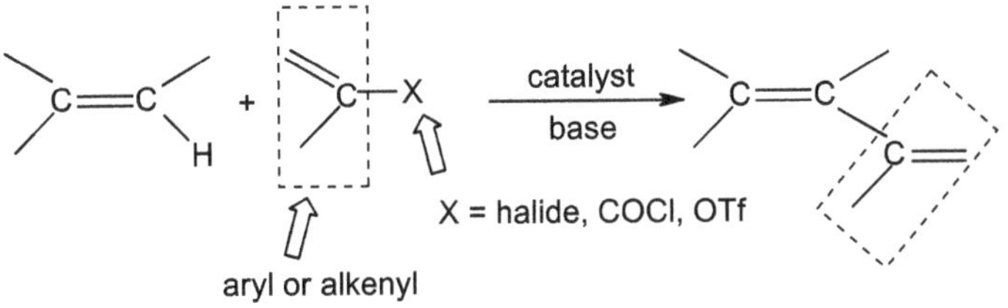

Scheme 3. Mizoroki–Heck reaction.

The first reported use of a thiosemicarbazone in the Heck reaction was in 2004 by the groups of Kovala-Demertzi and Kostas using a palladium complex of salicylaldehyde N(4)-ethylthiosemi carbazone (**1a**) (Figure 1) [15]. The crystal structure of the complex indicated that the ligand behaved as a tridentate ligand with N, S and O bonded to the metal. The reaction of styrene with a range of aryl bromides in the presence of varying concentrations of the complex was carried out in DMF (dimethylformamide) at 150 °C both in air and also under an argon atmosphere (Scheme 4). It was found that, as is normally the case for the Heck reaction, the catalytic activity was greater for aryl bromides with electron-withdrawing groups and decreased in the order NO_2 > CHO > H > OMe, leading to the conclusion that the oxidative addition of the aryl bromide to the complex was the rate-determining step. The use of an inert atmosphere in general gave better results, particularly for the least active aryl bromides and for low catalyst concentrations. However, the complex was stable enough in air under the reaction conditions to catalyse the reaction for the more activated aryl bromides, and turnover numbers (TONs) ranging from 120 to 14,000 and turnover frequencies (TOFs) in the range 5–583 h^{-1} were found. Using similar systems, **2**, involving derivatives of salicylaldehydethiosemicarbazone with an additional PPh_3 ligand, Xie et al. studied the catalysis of the Heck reaction of iodobenzene with methyl acrylate [94]. Having found that the methoxy-derivative **2c** gave the best yields in initial experiments, they examined the reaction with a range of other aryl iodides and aryl bromides, and with different solvents and bases. Generally, good to very good yields were obtained using aryl iodides and various acrylate esters under an argon atmosphere with DMF as solvent, K_2CO_3 as base, a temperature of 110–130 °C and a catalyst loading of at least 0.01 mol%. Using Na_2CO_3, lower catalyst loading also gave good results and this base was used in the reactions of the aryl bromides (Scheme 5). In the latter case, catalyst loadings of 0.1 or 1 mol% were necessary in order to obtain acceptable yields. Bidentate thiosemicarbazone complexes of palladium were investigated as catalysts in coupling reactions by Paul et al. [95]. They found that, in the Heck reaction, the complexes **3** and **4** displayed catalytic behavior at 0.5 mol% catalyst loading for the reaction between some aryl bromides and *n*-butyl acrylate (Scheme 6). The authors used Cs_2CO_3 as base and either ethanol-toluene or PEG (polyethylene glycol) as solvent at 110–150 °C. Although the results were only moderately good, this system could have much room for optimization taking into account the observation of Xie et al. (see above) that PEG was a poor solvent for the similar system that they examined and that K_2CO_3 was a superior base than Cs_2CO_3.

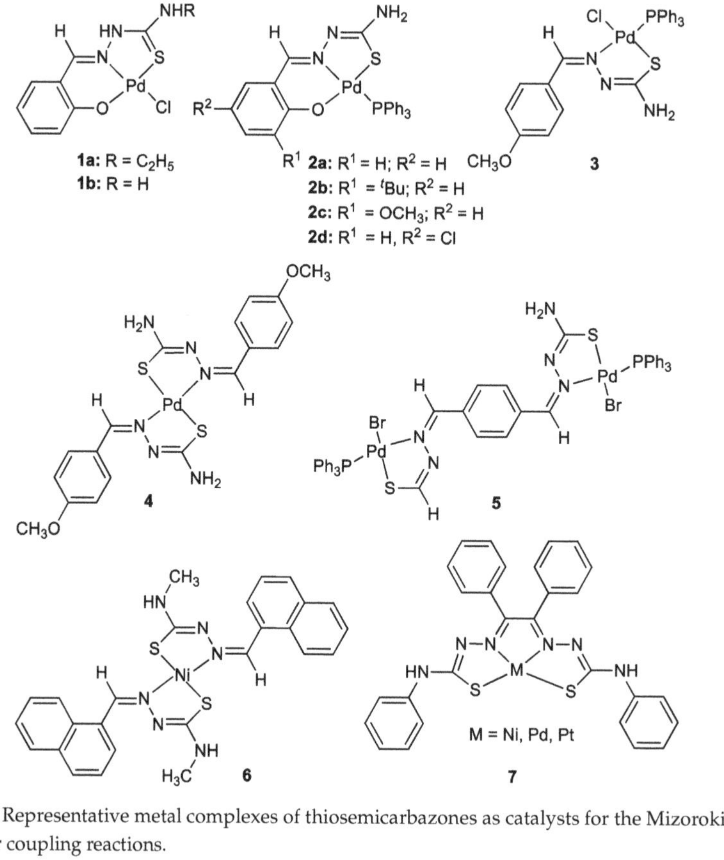

Figure 1. Representative metal complexes of thiosemicarbazones as catalysts for the Mizoroki–Heck and other coupling reactions.

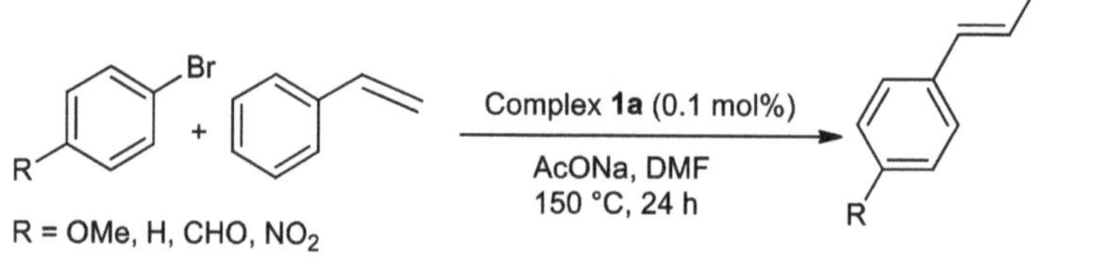

Scheme 4. Heck reaction of aryl bromides with styrene catalysed by complex **1a**: The first reported use of a thiosemicarbazone in the Heck reaction.

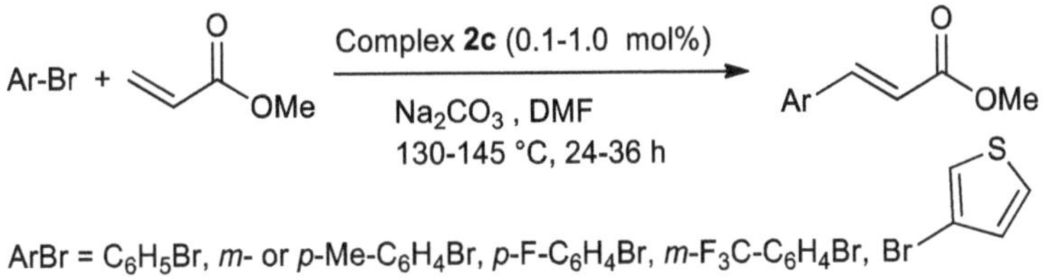

Scheme 5. Heck reaction of aryl bromides with methyl acrylate catalysed by complex 2c.

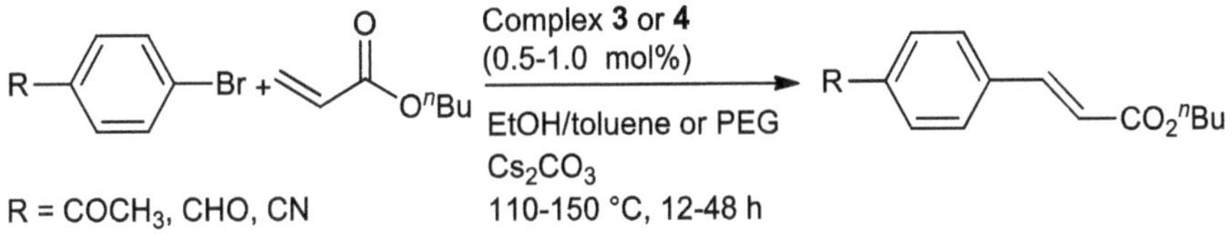

ArBr = C_6H_5Br, m- or p-Me-C_6H_4Br, p-F-C_6H_4Br, m-F_3C-C_6H_4Br, 2-bromothiophene

Scheme 6. Heck reaction of aryl bromides and n-butyl acrylate catalysed by complexes 3 or 4.

R = $COCH_3$, CHO, CN

The dinuclear bis-bidentate palladium complex 5, which also possesses PPh$_3$ ligands coordinated to the metal, was prepared and structurally characterized by Prabhu and Ramesh who subsequently made a systematic study of its catalytic activity in the Heck reaction of p-bromoacetophenone with t-butyl acrylate, examining the effect of temperature, solvent, base and catalyst loading [96]. Inorganic bases such as K_2CO_3 or Na_2CO_3 were superior to amines, DMF was the optimal solvent and a temperature of 100 °C provided the best results within a reasonable time. Catalyst loadings of 1 or 0.1 mol% gave quantitative yields but it is worth noting that the reaction proceeds even at very low catalyst loading of 0.00001 mol%, and, although the yield in this case is low (11%), the turnover number (1,100,000) and the turnover frequency (137,500 h^{-1}) are still impressive. Using optimized conditions, the authors were able to demonstrate the activity of the complex for a wide range of electron-withdrawing and electron-donating aryl bromides with methyl and t-butyl acrylate, styrene, p-methylstyrene, and p-chlorostyrene (Scheme 7). TONs of 6,000 to 9,800 and TOFs in the range 750–1225 h^{-1} were reported.

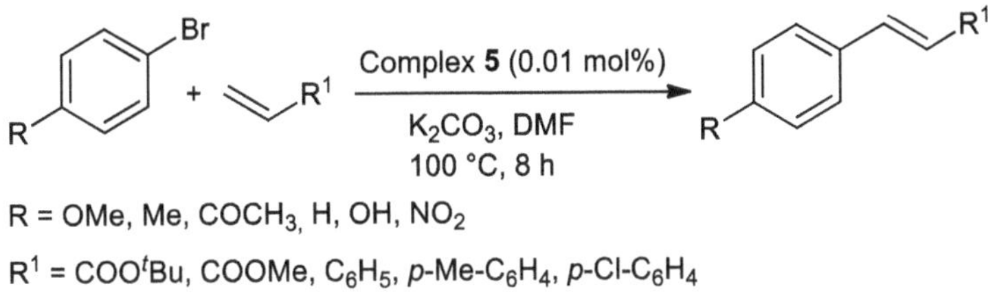

R = OMe, Me, $COCH_3$, H, OH, NO_2
R^1 = COO^{t}Bu, COOMe, C_6H_5, p-Me-C_6H_4, p-Cl-C_6H_4

Scheme 7. Heck reaction of aryl bromides with acrylate esters or substituted styrenes catalysed by complex 5.

The above studies all involved complexes of palladium but there have also been reports of the application of thiosemicarbazone nickel complexes to coupling reactions. One of the main motivations for this is the relatively low cost of nickel compared with palladium while, on the other hand, the main difficulty that needs to be surmounted is the well-established high efficiency of palladium complexes. It is also conceivable that there are important mechanistic differences in the mode of action of the complexes of the two metals but, for thiosemicarbazone complexes at least, no systematic studies have yet been carried out. The first report of the application of thiosemicarbazone nickel complexes to the Heck reaction was by Datta et al., who prepared and characterized three complexes with 2-hydroxyaryl thiosemicarbazone ligands [97]. Dinuclear complexes with tridentate N,S,O-coordination were formed which were reacted with PPh$_3$, pyridine, or bipyridine to give mononuclear complexes that retained the tridentate coordination. The complexes were examined for their catalytic efficiency in the reaction of p-bromoacetophenone, p-bromobenzonitrile, and p-bromobenzaldehyde with butyl acrylate in DMF at 130 °C. Catalyst loadings of 2 mol% were found to give good yields but TONs (18–50) and TOFs (2.1×10^{-4} sec^{-1}) were modest compared with TONs of analogous palladium complexes (8000). One encouraging feature, however, was that coupling reactions of aryl chlorides also proceeded with yields of a similar order of magnitude to the more reactive aryl bromides and iodides. Better results, at least as far as aryl bromides are concerned, were obtained using the nickel complex 6 reported by Suganthy et al. [98]. Using optimized conditions, this bis(thiosemicarbazone) nickel complex catalysed the reaction between a series of aryl bromides and methyl and t-butylacrylate, styrene, p-methylstyrene and p-chlorostyrene (Scheme 8). Using catalyst loadings of 0.5 mol%, moderate to very good conversions were obtained with turnover numbers ranging from 120 to 188 and turnover frequencies in the range 5–8 h^{-1}. However, it is significant that, compared with the system mentioned above [97], no catalytic activity was observed in the coupling of 4-chloroacetophenone with t-butyl acrylate in DMF/K$_2$CO$_3$ even after 24 h at elevated temperatures.

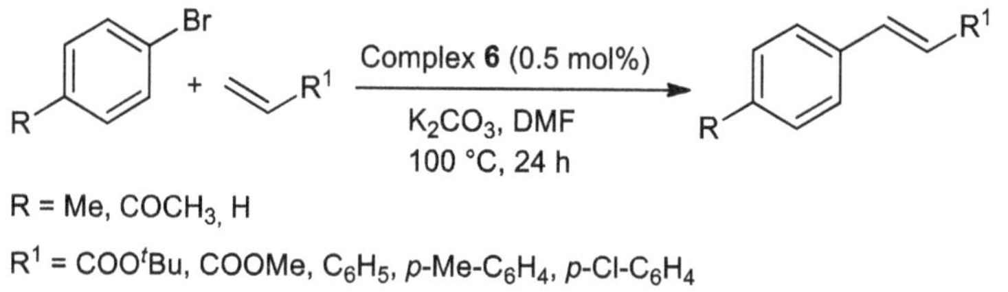

R = Me, COCH$_3$, H

R^1 = COO^{t}Bu, COOMe, C$_6$H$_5$, p-Me-C$_6$H$_4$, p-Cl-C$_6$H$_4$

Scheme 8. Heck reaction of aryl bromides with acrylate esters or substituted styrenes catalysed by Ni complex 6.

Very recently, a comparative study has been made of similar thiosemicarbazone complexes of nickel, palladium, and platinum [99]. Using a tetradentate bis-thiosemicarbazone ligand, Lima et al. synthesized the complexes 7. The tetradentate coordination was verified by X-ray diffraction structural determinations and the complexes were subsequently studied in the Heck reaction of iodobenzene with styrene. The palladium complex was an active catalyst at loadings of 3.5 mol% or above in a reaction carried out in DMF at 120 °C using triethylamine as the base. The platinum and nickel complexes showed activity but much less than the Pd complex. The palladium system appears to show much less catalytic activity than previously reported complexes and this may be due to the lack of free coordination sites in the tetracoordinated complex. On the other hand, it should be noted that the use of an organic base instead of an inorganic one is known to play a significant role and this also should be taken into account. It is also not clear from the report whether an inert atmosphere was employed. The authors performed preliminary DFT calculations from which they postulate that the process involving the tetradentate Pd complex does not follow the typical reaction mechanism for Heck catalysts involving an initial Pd(0)-Pd(II) oxidative-addition step. The calculations indicated a partial charge of +1.154 on the metal in the palladium complex compared to a much lower charge of +0.284 on the metal in the nickel one. Taking into account the increased catalytic activity of the Pd complex, and on the basis of their calculations for the likely steps in the catalytic cycle, the authors suggest that the reaction proceeds via an initial Pd(II)-Pd(IV) oxidative-addition of the aryl halide followed by olefin insertion and reductive elimination.

Published reports of the use of thiosemicarbazone complexes as described above for the Mizoroki–Heck reaction are summarised in Table 1 for indicative reactions.

Table 1. Mizoroki–Heck reactions catalysed by thiosemicarbazone complexes: representative conditions and yields [1].

Metal	T (°C)	Solvent	Time (h)	Ligand [2]	Base	Catalyst (mol%)	Yield (%)	Ref.
Pd	150	DMF	24	O,N,S	NaOAc	0.1	46–95	[15]
Pd	130–145	DMF	24–36	O,N,S	Na_2CO_3	0.1–1.0	50–90	[94]
Pd	110–150	EtOH/toluene or PEG	12–48	N,S	Cs_2CO_3	0.5–1.0	57–80	[95]
Pd	100	DMF	8	N,S	K_2CO_3	0.01	60–97	[96]
Ni	130	DMF	24	O,N,S	Cs_2CO_3	2.0	36–99	[97]
Ni	110	DMF	24	N,S	K_2CO_3	0.5	60–94	[98]
Pd	120	DMF	5–24	S,N,N,S	Et_3N	3.5	67–82 [3]	[99]

[1] conditions refer to reactions involving aryl bromides and (substituted) styrenes or acrylates. [2] ligand donor atoms. [3] for the reaction of PhI with styrene.

2.2. Suzuki–Miyaura and Related Reactions

The coupling of alkenyl, alkynyl, and aryl halides with boronic acids and related derivatives by palladium complexes, first reported in 1979 by Miyaura and Suzuki, led to intense research activity aiming at optimising the reaction and extending its application to increasingly more demanding systems (Scheme 9). Numerous reviews have appeared and continue to appear on the subject [2,3]. Representative metal complexes of thiosemicarbazones that have been used as catalysts for this reaction are shown in Figures 2–4; see also in Figure 1.

The first report of the use of thiosemicarbazone complexes in this reaction was by Kostas et al., who studied the cross-coupling of aryl halides with phenylboronic acid (Scheme 10) [16]. The complexes **1a** and **1b** used were derived from salicylaldehyde (Figure 1), and one of them (**1a**) having already been successfully used in the Heck reaction as mentioned above [15]. The complexes are air-stable and this therefore enabled the reactions to be carried out without the need for an inert atmosphere. In addition, they are moisture-stable and in fact the addition of one equivalent of water was found to be beneficial. Aryl bromides with varying substitution were used and, as had been previously observed in reactions with other catalysts, the best results were obtained with electron-withdrawing substituents. Catalyst loadings of 0.1 mol% gave moderate to very good conversions for most substrates in reactions in DMF at 100 °C using Na_2CO_3 as base, but even lower loadings of 0.001 mol% were also active systems, albeit with lower conversions. TONs ranging from 400 to 49,000 and TOFs in the range 17–2042 h^{-1} were reported. The reaction with aryl chlorides was also catalysed by these complexes but, as expected, with somewhat lower TONs (260–370) and TOFs (11–15 h^{-1}). These complexes were less active than some previously reported P-systems but have the important advantage of being phosphane free and requiring less demanding conditions. In a subsequent paper, the same group reported on the synthesis and characterisation of a new palladium thiosemicarbazone complex **8** (Figure 2) [100]. The thiosemicarbazone was again derived from salicylaldehyde but with a tertiary amino end group derived from hexamethyleneimine. The crystal structure determination of the bis-ligand palladium complex demonstrated that the two ligands were coordinated in a bidentate fashion via N and S donors. In contrast with the complexes in the previous study, the oxygen on the salicylaldehyde portion was not involved in direct bonding to the metal but was connected via a hydrogen bond to the unsubstituted thioamide nitrogen. This complex did not demonstrate catalytic activity in the Suzuki–Miyaura reaction using the conditions employed in the previous work, and it was proposed that this is because the metal is bonded to two thiosemicarbazone moieties by four intramolecular bonds, resulting in inhibition of the addition of the aryl halide to the metal during the catalytic cycle. However, using microwave irradiation, positive results were obtained for the reaction of bromobenzene and p-nitrobromobenzene with phenylboronic acid (Scheme 11). As in the previous study, the addition of water was found to be beneficial and good yields were obtained after up to 60 min irradiation using DMF as solvent and Na_2CO_3 as base. Catalyst loadings of 0.1 mol% were used for reactions with bromobenzene and 0.001 mol% for those with p-nitrobromobenzene. TONs of up to 37,000 and TOFs of up to 617 min^{-1} were recorded. The reaction of phenylboronic acid with *p*-nitrochlorobenzene was also successful under these conditions. The observation that conventional heating failed to promote the reaction prompted the authors to postulate that the acceleration of the reaction was due to specific microwave effects [100,101].

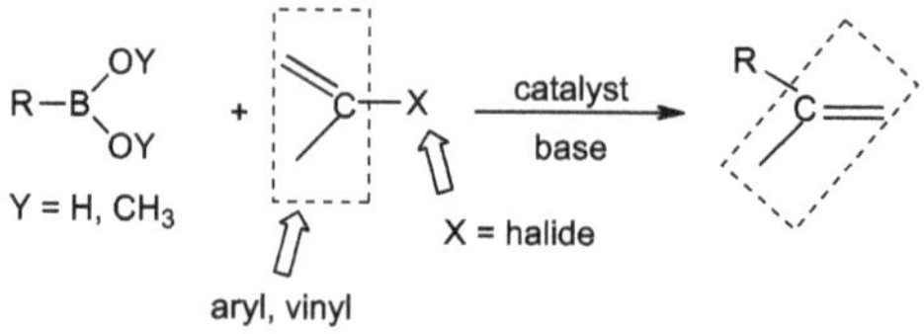

Scheme 9. Suzuki–Miyaura reaction.

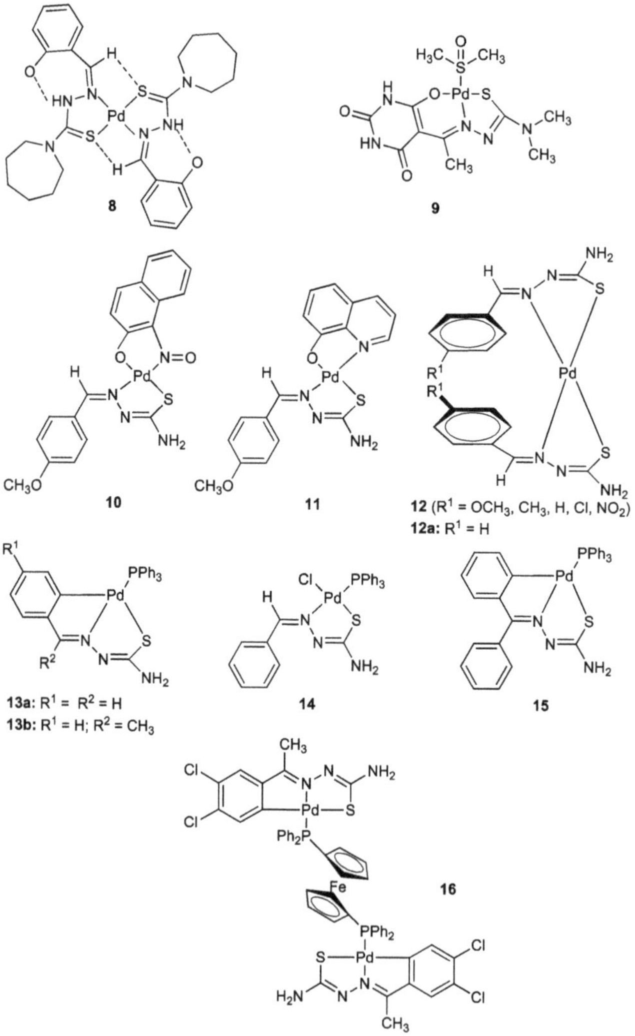

Figure 2. Representative metal complexes of thiosemicarbazones as catalysts for the Suzuki–Miyaura and other coupling reactions (Part A).

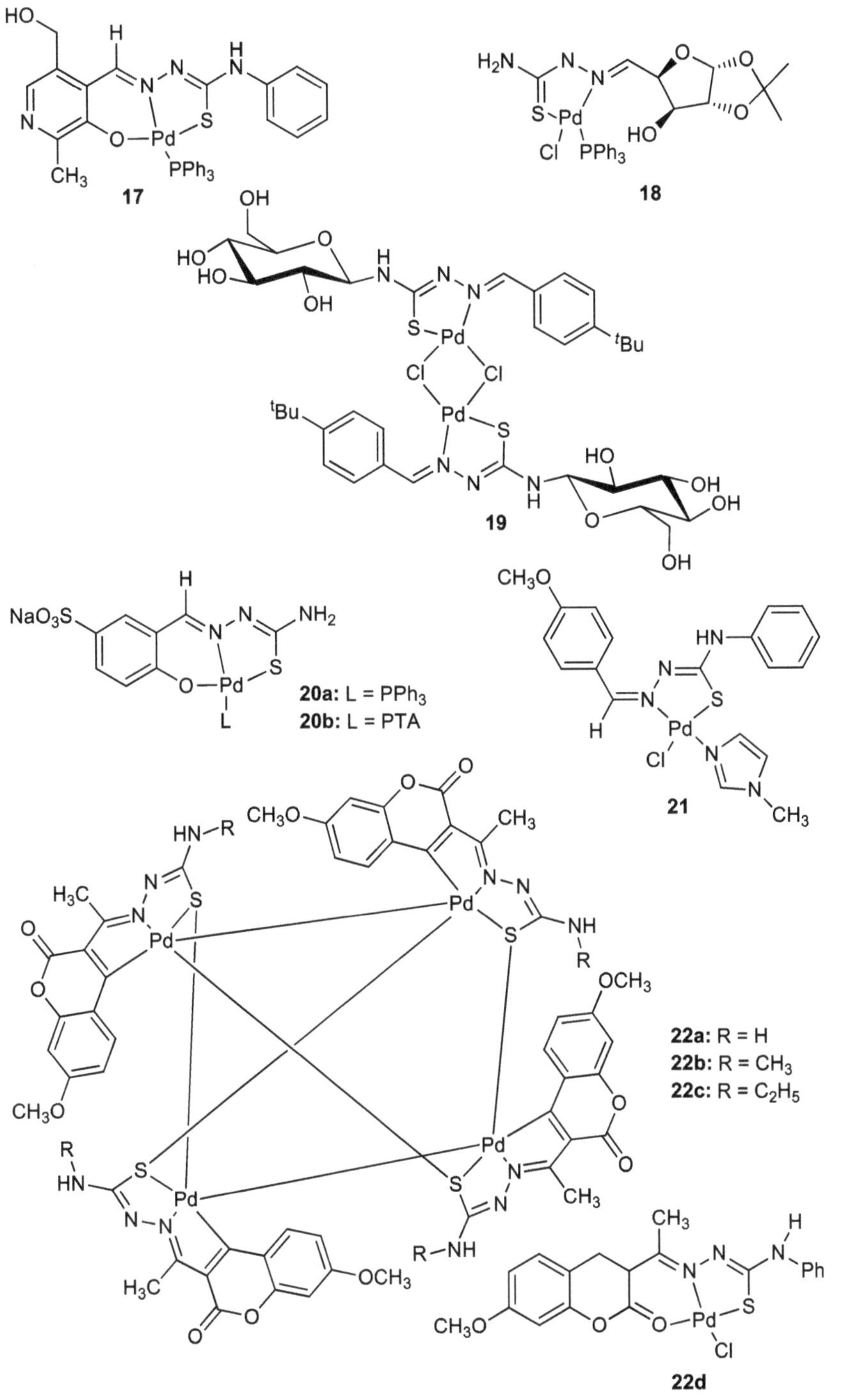

Figure 3. Representative metal complexes of thiosemicarbazones as catalysts for the Suzuki–Miyaura and other coupling reactions (Part B).

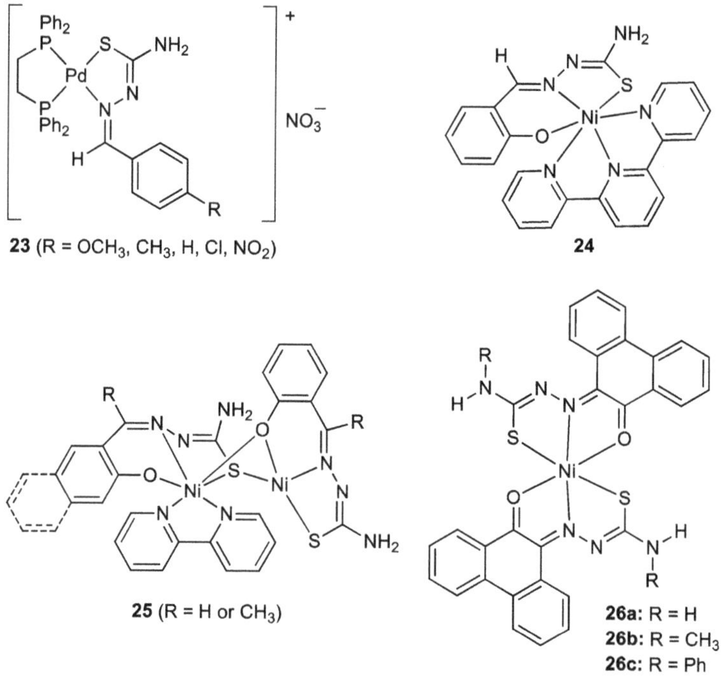

23 (R = OCH₃, CH₃, H, Cl, NO₂)

25 (R = H or CH₃)

26a: R = H
26b: R = CH₃
26c: R = Ph

Figure 4. Representative metal complexes of thiosemicarbazones as catalysts for the Suzuki–Miyaura and other coupling reactions (Part C).

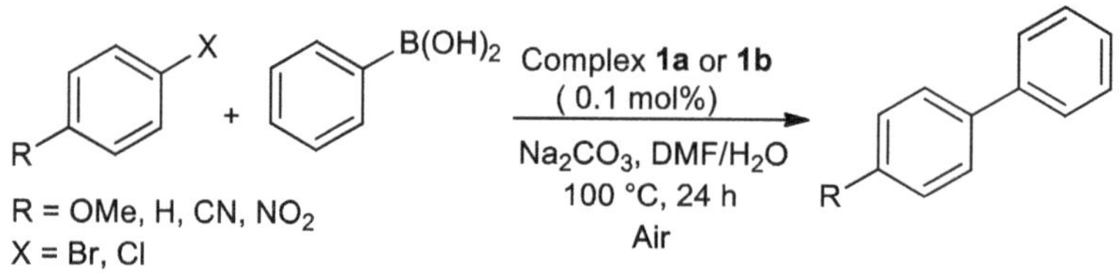

Scheme 10. Suzuki reaction of aryl halides with phenylboronic acid catalysed by complex **1a** or **1b**: The first reported use of a thiosemicarbazone in the Suzuki reaction.

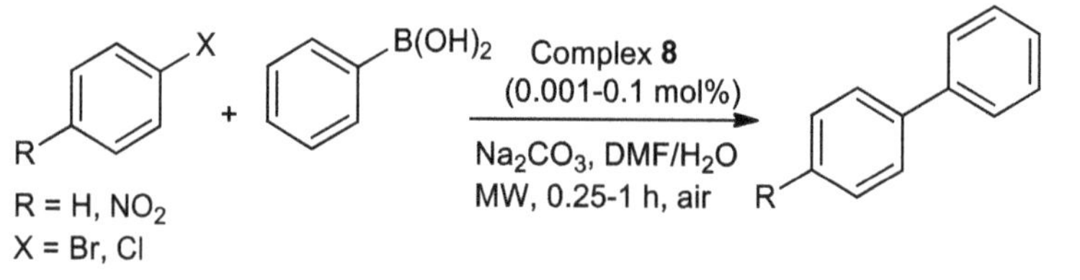

Scheme 11. Microwave-promoted Suzuki–Miyaura cross-coupling of aryl halides with phenylboronic acid catalysed by complex **8**.

Bidentate complexes **3** and **4** prepared and characterised by Paul et al., and described above in relation to their catalytic activity in the Heck reaction (Figure 1), were also examined for their possible use in the Suzuki–Miyaura reaction [95,102]. The coupling of phenylboronic acid with p-bromoacetophenone, p-bromobenzaldehyde, or p-bromobenzonitrile was investigated using both the mono-ligand complex **3** containing PPh₃ as a supporting ligand and the bis-ligand phosphane-free complex 4 (Scheme 12). Relatively mild conditions were used (ethanol-toluene or PEG solvent, NaOH as base, temperature 95–110 °C) and very high TONs (100,000) and TOFs (up to 11,111 h^{-1}) were recorded for the former complex while lower, but still high, TONs (up to 8,800) and TOFs (up to 733 h^{-1}) were recorded in the latter case. The mono-ligand complexes showed tolerance to water although the best results were obtained in dry conditions. The reaction involving p-bromoacetophenone using the mono-ligand complex 3 was even successful at 25 °C giving a conversion of 99%, TON of 100,000, and TOF of 5,000 h^{-1}.

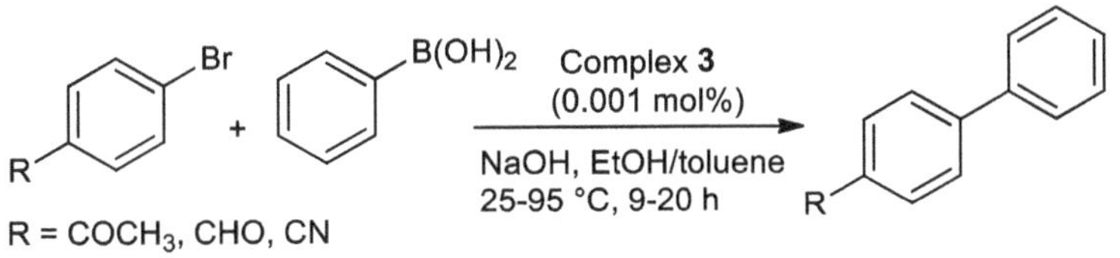

Scheme 12. Suzuki reaction of aryl bromides with phenylboronic acid catalysed by complex 3.

The phosphane-free complex 9 was reported by Castiñeiras et al. (Figure 2) [103]. The tridentate coordination of the dianionic ligand derived from 5-acetylbarbituric-4N-dimethylthiosemicarbazone was confirmed by XRD crystallography. In the reaction between phenylboronic acid and bromobenzene, p-bromoanisole, p-bromonitrobenzene and the corresponding chloro-derivatives, conversions of between 46 and 78% were observed for the aryl bromides, while somewhat lower values from 21 to 32% were found for the chlorides (Scheme 13). The authors postulated that, since the ligand is dianionic, the mechanism involves initial oxidative addition of the aryl halide and cycling between palladium +2 and +4 oxidation states as had previously been proposed for other Pd catalysts possessing pincer ligands, rather than via the 0 and +2 states [104,105]. However, it should be noted that it is now generally accepted that cross-couplings catalysed by cyclometallated Pd(II) complexes proceed via a Pd(II) to Pd(0) pathway and that Pd(0) species are the active catalysts [57].

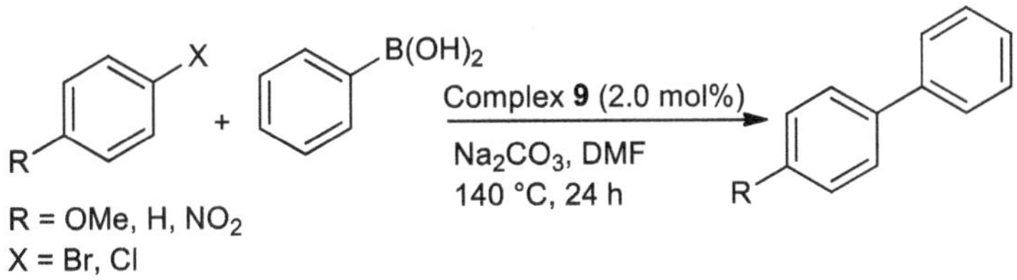

Scheme 13. Suzuki reaction of aryl halides with phenylboronic acid catalysed by complex 9.

In 2012 the group of Bhattacharya reported new thiosemicarbazone complexes (10, 11) of palladium with 1-nitroso-2-naphtholate or quinolin-8-olate supporting ligands (Figure 2) [106]. The thiosemicarbazone ligands were derivatives of benzaldehyde with a range of para-substituents in order to investigate their effect on catalytic activity. The complexes contain two 5-membered rings and only the configuration where the two nitrogens are trans to each other was observed. Catalysis of the Suzuki–Miyaura reaction by complexes 10 and 11 was studied, with optimisation of certain parameters being carried out with phenylboronic acid and p-bromoacetophenone as substrates. Using catalyst loadings of 0.001 mol%, 100% conversions were obtained after 24 h in PEG at 120 °C using either NaOH or Cs$_2$CO$_3$ as base (Scheme 14). Under these conditions, TONs of up to 100,000 and TOFs of up to 16,667 h^{-1} were recorded. Reduction of the loading to 0.0001 mol% gave slightly lower conversions as did replacement of p-bromoacetophenone with the less reactive p-bromobenzaldehyde or p-bromobenzonitrile. Other halo-derivatives were also examined and, as expected, p-iodoacetophenone also gave 100% conversion while p-chloroacetophenone required a higher loading of 0.1 mol% to achieve a similar result. Interestingly, low but significant yields, 10–12%, of the product of coupling p-fluoroacetophenone with phenylboronic acid were observed with 1 mol% catalyst loading. Although no specific studies of the likely mechanism were described, the authors favoured a process involving initial formation of a zerovalent Pd species, in which the protonated thiosemicarbazone and N,O-donor ligands remain coordinated, followed by oxidative addition of the aryl halide. If indeed the active species is as proposed, it could be of interest to determine if it remains active for reuse in repeated cycles. In an attempt to produce analogous complexes with 2-picolinic acid as the supporting ligand, Dutta and Bhattacharya, instead of the expected mono-ligand complexes, obtained the bis-ligand complexes **12** with a rare cis-configuration and also a second product which was postulated to be a polymeric bridged complex containing a tridentate cyclometallated ligand (Figure 2) [107]. This was confirmed by cleavage of the bridges by triphenylphosphine to give the mononuclear complexes of the type **13a** whose structures were also confirmed by X-ray crystallography. The two sets of complexes were examined for their potential as catalysts for the Suzuki–Miyaura reaction for a range of aryl halides and substituted phenylboronic acids (Scheme 15). Very good conversions were observed for most of the reactions with aryl iodides and bromides under relatively mild conditions (PEG as solvent, 120 °C, NaOH as base, 1–8 h) and catalyst loadings of 0.001 mol%, while aryl chlorides required higher catalyst loadings of 0.1 mol% to achieve comparable results. The p-methoxyphenyl and p-chlorophenylboronic acids reacted more sluggishly than phenylboronic acid itself. Of the two sets of complexes, the mono-ligand complexes gave somewhat superior results and this was attributed to the presence of the triphenyphosphine supporting ligand. For both sets of complexes, no additional ligand was needed and the authors argue that this implies that the ligands in the pre-catalyst do not dissociate and that they stabilize the intermediate Pd(0) species. The same research group has also reported further examples of the mono-ligand cyclometallated complex with PPh_3 supporting ligand (complexes **13b**, **14**, **15**). These were prepared by a slightly different route and, together with a non-cyclometallated complex containing a bidentate thiosemicarbazone ligand as well as PPh_3, were examined for their activity in the coupling of p-haloacetophenones with phenylboronic acid (Scheme 16) [108]. Results similar to those given above were obtained, the cyclometalated complexes giving the better results. Notably, coupling of the fluoro-derivative could also be achieved with these catalysts. Analogous cyclometallated palladium complexes **16** based on 3,4-dichloroacetophenone thiosemicarbazone have also been reported by Yan et al. [109]. These complexes were screened for their activity in the Suzuki–Miyaura reaction and the most promising of the four, a dinuclear complex with a 1,1′-bisdiphenylphosphinoferrocene bridging supporting ligand, was used for further study. Reactions were carried out for 24–48 h in air or argon, using DMF as a solvent, K_3PO_4 as base and a temperature of 130 °C, using a range of aryl bromides and chlorides and various aryl boronic acids (Scheme 17). Substitution on the boronic acid had no major effect except for 2-methoxyphenyl boronic acid, which gave lower yields, possibly because of steric effects. The aryl bromides all gave moderately good to excellent yields while the chlorides, as expected, gave lower conversions except for p-nitrochlorobenzene.

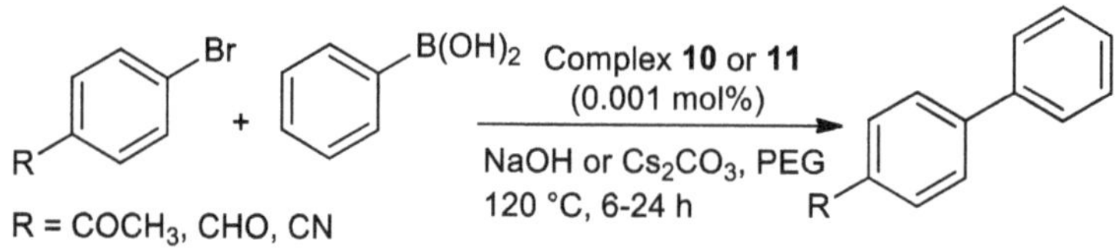

Scheme 14. Suzuki reaction of aryl bromides with phenylboronic acid catalysed by complex **10** or **11**.

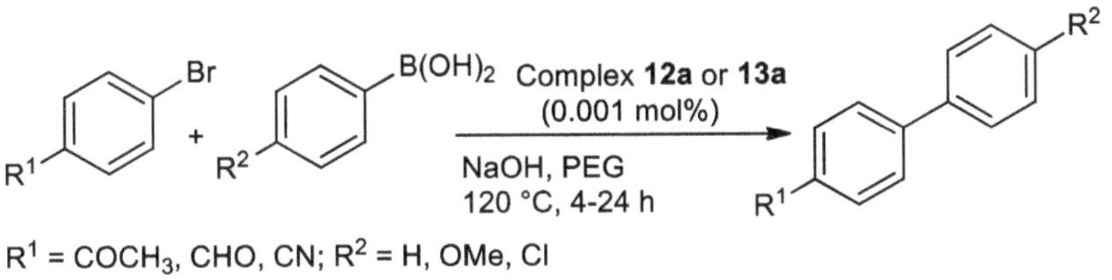
Br
B(OH)2
Complex 12a or 13a
(0.001 mol%)
NaOH, PEG
120 °C, 4-24 h
R1
R2
R1 = COCH3, CHO, CN; R2 = H, OMe, Cl

Scheme 15. Suzuki reaction of aryl bromides with substituted phenylboronic acids catalysed by complex **12a** or **13a**.

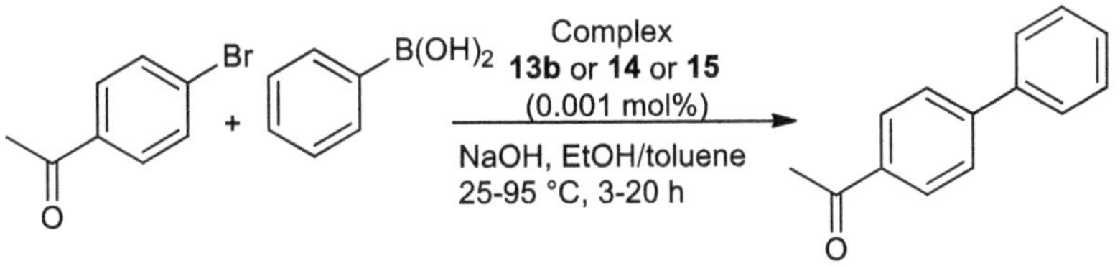
Br
B(OH)2
Complex
13b or 14 or 15
(0.001 mol%)
NaOH, EtOH/toluene
25-95 °C, 3-20 h
O

Scheme 16. Suzuki reaction of *p*-bromo-acetophenone with phenylboronic acid catalysed by complex **13b**, **14** or **15**.

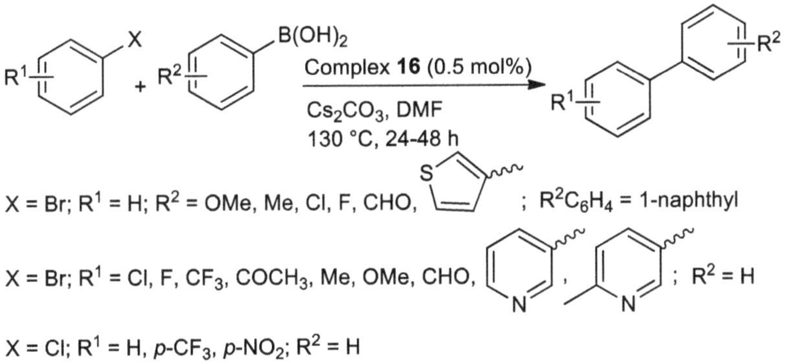
X
B(OH)2
Complex 16 (0.5 mol%)
Cs2CO3, DMF
130 °C, 24-48 h
R1
R2
X = Br; R1 = H; R2 = OMe, Me, Cl, F, CHO,
; R2C6H4 = 1-naphthyl
X = Br; R1 = Cl, F, CF3, COCH3, Me, OMe, CHO,
; R2 = H
X = Cl; R1 = H, p-CF3, p-NO2; R2 = H

Scheme 17. Suzuki reaction of aryl halides with substituted phenylboronic acids catalysed by complex **16**.

The complex **17** was reported by Pandiarajan et al. (Figure 3) [110]. The dianionic ligand binds through S, N, and O donors and the complex is air and moisture stable. It catalysed the Suzuki–Miyaura reaction in refluxing DMF with K_2CO_3 base with very good conversions after 3 h for a number of aryl bromides and boronic acids (Scheme 18). Coupling of *p*-iodoacetophenone was also achieved with excellent conversion while with the analogous chloro-derivative moderate yields of product were obtained after 12 h. In the same year, another phosphane supported thiosemicarbazone palladium complex, **18**, was reported by Verma et al. (Figure 3) [111]. The thiosemicarbazone in this case is derived from a sugar aldehyde and the complex from the analogous semicarbazone was also prepared. These ligands were shown by structural studies to bind to the metal in a bidentate manner. The authors were particularly interested in the catalytic activity of these complexes in the coupling of aryl chlorides with boronic acids and found that this could be achieved in good to excellent yields at ambient temperatures using a catalyst loading of 0.2 mol% in EtOH, with K_2CO_3 as base and reaction times of just 30–90 min (Scheme 19). At lower catalyst loadings, however, the reaction times were much longer and yields were also reduced. The authors were able to demonstrate that the catalysts retained their activity after five cycles.

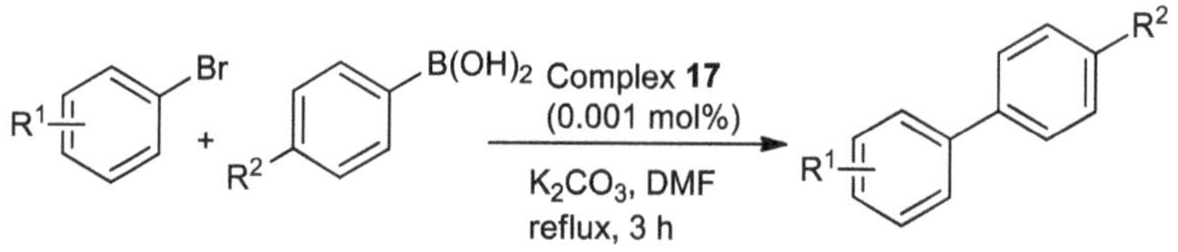

R^1 = p-COCH$_3$, o-, p-Me, p-OMe; R^2 = H, Cl, Me, OMe

Scheme 18. Suzuki reaction of aryl bromides with substituted phenylboronic acids catalysed by complex **17**.

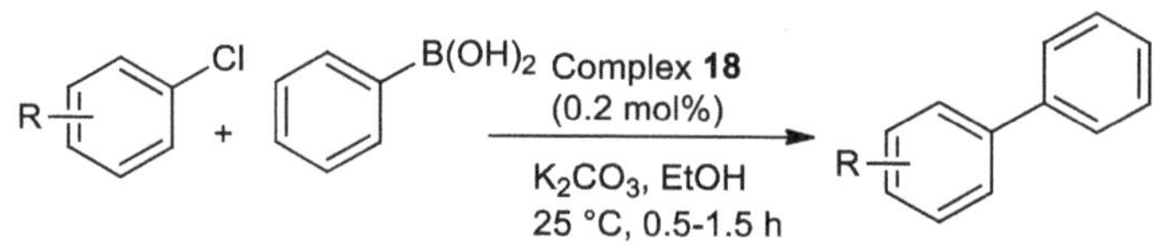

R = o-, p-Cl, p-CN, p-COCH$_3$, p-NH$_2$, p-CHO, p-NO$_2$

Scheme 19. Suzuki reaction of aryl chlorides with phenylboronic acid catalysed by complex **18**.

In an attempt to develop a phosphane-free catalyst, the group of Kostas synthesised a binuclear palladium complex **19** with a ligand derived from β-D-glucopyranosyl-thiosemicarbazone (Figure 3) [112]. The complex was characterised spectroscopically and investigated as a potential catalyst for the Suzuki–Miyaura reaction between aryl bromides and phenyl boronic acid (Scheme 20). After 24 h at 100 °C in DMF and with K$_2$CO$_3$ as base, good to excellent conversions were obtained using a 0.05 mol% catalyst loading. Aryl chlorides, however, gave rather poor conversions. Tests were carried out in order to determine the nature of the catalyst and it was concluded that the active species was heterogeneous and possibly composed of Pd(0) nanoparticles. The complexes used in this study and in the study mentioned in the previous paragraph [111] are of additional interest in that they employ chiral ligands. Although in these studies, possible applications in asymmetric catalysis were not explored, the amenability of thiosemicarbazone ligands to functionalisation with chiral groups could provide a promising avenue for future work.

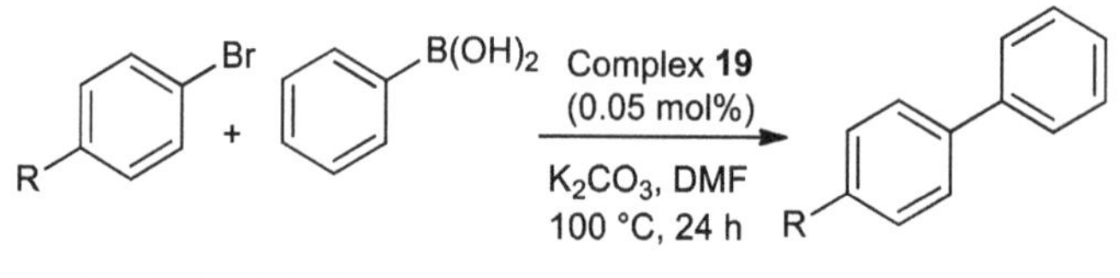

R = Me, OMe, H, PhCO, NO$_2$, CN, CHO; RC$_6$H$_4$ = 1-naphthyl

Scheme 20. Suzuki reaction of aryl bromides with phenylboronic acid catalysed by complex **19**.

The use of aqueous media for carrying out catalytic reactions has many attractions and in 2017, Matsinha et al. reported the synthesis of two water-soluble palladium complexes **20a** and **20b** containing sulfonated-thiosemicarbazone ligands (Figure 3) [113]. In both complexes, the ligand is tridentate, and the vacant position is occupied by a tertiary phosphine (PPh$_3$ in **20a** and 1,3,5-triaza-7-phosphaadamantane in **20b**). The complexes displayed good stability in water. Catalytic coupling of a range of aryl bromides with aryl boronic acids was investigated in water at 70 °C using Na$_2$CO$_3$ as base and TBAB (tetrabutylammonium bromide) as a phase-transfer mediator (Scheme 21). Satisfactory results were obtained, although it should be noted that rather long reaction times (24 h) and higher catalyst loadings (1 mol%) were employed than were usual for reactions in non-aqueous media. An investigation into the reusability of the catalysts indicated that activity drops off quite rapidly and that during the fourth cycle activity was low. The authors speculate that this could be due either to leaching of the active catalyst during the extraction step or to partial decomposition of the active species. However, the possibility that the catalyst was the precursor to a heterogeneous system which then degraded quickly was ruled out by the authors, since the mercury drop test for such cases failed to affect the catalytic activity to any significant extent. An aqueous media was also employed by Baruah et al. for the complex **21** (Figure 3) [114]. In this case, the supporting ligand is imidazole and the thiosemicarbazone adopts bidentate coordination as a monoanion. After a number of optimisation runs, the authors examined the coupling of a range of aryl halides and aryl boronic acids using this complex as a precatalyst. Ambient temperatures were employed with K_2CO_3 as the base and a catalyst loading of 1.18 mol%. For most of the aryl bromides, good conversions were achieved after 2–6 h, while the aryl chorides examined needed an elevated temperature (60 °C) and longer reaction times for comparable results. The complex itself was not soluble in water and was used as a suspension and it was suspected that the actual catalyst could be a Pd(0) species. Support for this came from a mercury drop test, which inhibited catalytic activity. The activity of the catalyst falls of in subsequent cycles but no significant leaching of palladium was observed. The catalyst isolated after a first cycle was therefore examined by TEM, SEM-EDX, and XRD and was determined to consist of Pd(0) nanoparticles whish are presumed to be formed by dissociation of the ligands from the initial complex during the reaction. SEM-EDX examination of these nanoparticles indicated that they are possibly stabilized by surface thiosemicarbazone ligands. They were found to have an initial size of 1.5–2.0 nm, but after successive runs they aggregated to larger particles with lower activity.

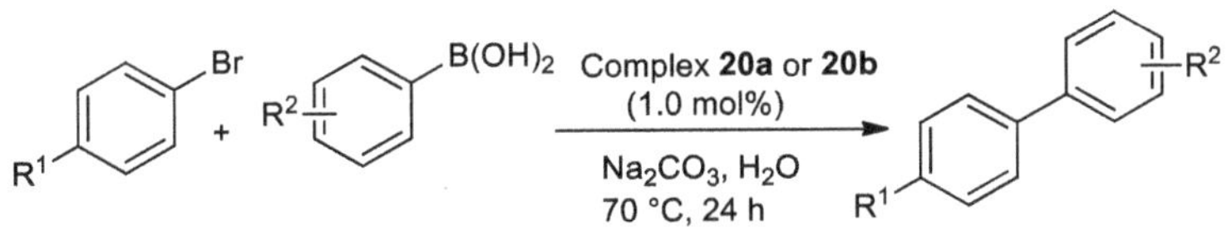

Scheme 21. Aqueous Suzuki reaction of aryl bromides with substituted phenylboronic acids catalysed by complex **20a** or **20b**.

Dharani et al. reported a series of palladium complexes **22** derived from 3-acetyl-7-methoxy-2H-chromen-2-one thiosemicarbazones (Figure 3) [115]. Three of the products (**22a–c**) proved to be tetranuclear complexes in which ligands are bonded via S, N, and C, cyclometallation having taken place by activation of the ortho-C–H bond. The palladium atoms are connected via thiolate bridges. The fourth complex, with phenyl substitution on the terminal nitrogen of the thiosemicarbazide, was the mononuclear species **22d**. All of the complexes were screened for activity as catalysts for the Suzuki–Miyaura reaction and one them (**22b**) was chosen for further study. Using a 0.125 mol% loading of the complex, EtOH-H_2O as solvent, K_2CO_3 as base, and a temperature of 70 °C, good conversions were obtained for the coupling of phenyl boronic acid with a range of aryl halides including chloroquinolines. The results were found to compare well with those obtained for other tetranuclear palladium complexes in aqueous conditions. The catalyst isolated from the reaction could be used up to four more times with only partial loss of activity. In further cycles, however, a 50% loss of activity occurred. A mechanism was proposed involving initial cleavage of the tetranuclear complex into a mononuclear species followed by a Pd(II)-Pd(IV) oxidative addition/elimination sequence. The fall-off in activity in fifth and successive cycles was ascribed to the gradual aggregation of the mononuclear species to form less active nanoparticles, evidence for which was obtained by powder X-ray diffraction studies. More recently, Bakir et al. have reported similar tetranuclear complexes derived from di-thienyl ketone thiosemicarbazone [116]. These were screened for their possible use as precatalysts in the Suzuki–Miyaura reaction but the results were only moderate. This was ascribed by the authors to be at least partly due to the polymeric nature and insolubility of the complex.

Cationic complexes **23** of the type [Pd(dppe)L]NO$_3$ (dppe = 1,2-bis(diphenylphosphino)ethane), where L is a bidentate thiosemicarbazone ligand derived from a *p*-substituted benzaldehyde were prepared by Thapa et al., and structurally characterised, confirming the formation of *N,S*-chelated 5-membered rings (Figure 4) [117]. The authors hypothesised that, in view of the previously observed improvements in catalytic efficiency due to the presence of phosphine supporting ligands, the use of a diphosphine could potentially enhance this even further. Indeed, compared with other analogous complexes prepared by these workers [95], superior results were seen. Good conversions with high TONs (up to 980,000) and TOFs (up to 326,667 h^{-1}) were observed for a number of aryl iodides and bromides at 95 °C in EtOH-toluene with Cs$_2$CO$_3$ as base and catalyst loadings of 0.001–0.0001 mol% (Scheme 22). Chlorides also engaged quite readily in the coupling reaction with somewhat higher catalyst loadings and slightly modified conditions, while aryl fluorides could also be coupled with the unsubstituted phenyl boronic acid at 130 °C in PEG using NaOBu^t as base and with a 1 mol% catalyst loading.

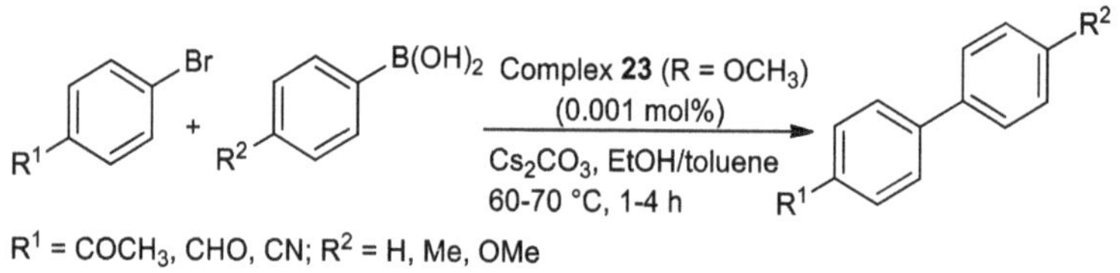

Scheme 22. Suzuki reaction of aryl bromides with substituted phenylboronic acids catalysed by complex **23**.

Catalysis of the Suzuki–Miyaura reaction by nickel complexes has attracted attention in recent years due to the greater accessibility of nickel and also its greater activity in certain cases. Thiosemicarbazone complexes of nickel, however, have been much less investigated than their palladium counterparts. In 2011, Datta et al. reported the synthesis of mono- and dinuclear nickel complexes **24** and **25**, respectively, derived from salicylaldehyde, 2-hydroxyacetophenone and 2-hydroxynaphthaldehyde thiosemicarbazones with bipyridine or terpyridine supporting ligands (Figure 4) [118]. The tridentate ligands are bonded via N, S, and O donors and the dinuclear complexes are bridged via thiolate and phenolate groups. The complexes were examined for their activity in the Suzuki–Miyaura reaction for some aryl bromides and iodides with phenyl boronic acid. Relatively good activity was observed although it was much less than that shown by similar palladium complexes. Similar *O,N,S*-bonded nickel complexes **26** derived from 9,10-phenanthrenequinone thiosemicarbazone, 9,10-phenanthrenequinone *N*-methylthiosemicarbazone and 9,10-phenanthrenequinone *N*-phenylthiosemicarbazone have also been reported by Anitha et al., but these complexes gave only rather moderate results for Suzuki–Miyaura couplings [119].

Although it is not a normal Suzuki–Miyaura reaction, we may also mention here the application of the palladium thiosemicarbazonato complex **27** as a catalyst for the synthesis of diaryl ketones via the C–C coupling reaction between aryl aldehydes and aryl boronic acids reported by Prabhu and Ramesh (Scheme 23) [120]. Optimal conditions were found to be 110 °C in toluene in the presence of Cs$_2$O$_3$ and using 5 mol% of the complex. The scope of the reaction was demonstrated by the synthesis of diaryl ketones from the reaction of a wide variety of aromatic and heteroaromatic aldehydes with phenyl boronic acid as well as from the reaction of a selection of aryl boronic acids with benzaldehyde. Satisfactory to excellent isolated yields were obtained.

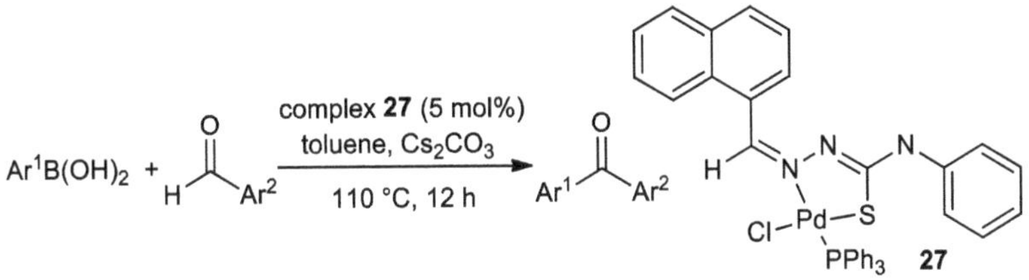

Scheme 23. Synthesis of diaryl ketones by carbon–carbon coupling reaction between aryl aldehydes and aryl boronic acids.

Table 2 summarises representative conditions and yields for Suzuki–Miyaura reactions catalysed by thiosemicarbazone complexes.

Table 2. Suzuki–Miyaura reactions catalysed by thiosemicarbazone complexes: representative conditions and yields [1].

Metal	T (°C)	Solvent	Time (h)	Ligand [2]	Base	Catalyst (mol%)	Yield (%)	Ref.
Pd	100	DMF/H_2O	24	O,N,S	Na_2CO_3	0.1	40–88	[16]
Pd	100–157	DMF/H_2O	0.25–1	O,N,S	Na_2CO_3	0.001–0.1	25–85	[100] [3]
Pd	25–95	EtOH/toluene	9–20	N,S	NaOH	0.001	>99	[102]
Pd	140	DMF	24	O,N,S	K_2CO_3	2.0	46–78	[103]
Pd	120	PEG	6–24	N,S	NaOH or Cs_2CO_3	0.001	100	[106]
Pd	120	PEG	6–24	N,S	NaOH	0.001	65–100	[107]
Pd	120	PEG	4–24	C,N,S	NaOH	0.001	71–100	[107]
Pd	25–95	EtOH/toluene	9–20	N,S	NaOH	0.001	98–100	[108]
Pd	25–95	EtOH/toluene	3–14	C,N,S	NaOH	0.001	100	[108]
Pd	130	DMF	24–48	C,N,S	K_3PO_4	0.5	31–99	[109]
Pd	reflux	DMF	3	O,N,S	K_2CO_3	0.001	78–99	[110]
Pd	25	EtOH	0.5–1.5	N,S	K_2CO_3	0.2	76–98	[111] [4]
Pd	100	DMF	24	N,S	K_2CO_3	0.05	60–99	[112]
Pd	70	H_2O	24	O,N,S	Na_2CO_3	1.0	25–98	[113]
Pd	28	H_2O	2–12	N,S	K_2CO_3	1.18	65–90	[114]
Pd	60–70	EtOH/H_2O	1–4	C,N,S	K_2CO_3	0.125	51–99	[115]
Pd	95	EtOH/toluene	6–8	N,S	Cs_2CO_3	0.001	79–100	[117]
Ni	140	DMF	24	O,N,S	Cs_2CO_3	2.0	40–99	[118]
Ni	90	DMA	7	O,N,S	K_2CO_3	1.0	28–64	[119]
Pd	110	toluene	12	N,S	Cs_2CO_3	5.0	62–97	[120] [5]

[1] conditions refer to reactions involving aryl bromides and phenyl or aryl boronic acids. [2] ligand donor atoms. [3] microwave irradiation [4] aryl chlorides were used. [5] Aryl aldehydes used instead of aryl halides.

2.3. Sonogashira and Related Reactions

Since the first report by Sonogashira in 1975 [121], the metal complex-catalysed coupling of terminal alkynes with haloorganics has developed into an essential tool for the synthetic organic chemist (Scheme 24) [1,5,9]. Palladium or copper complexes are generally employed to facilitate this reaction and some very efficient systems have been reported for a wide variety of halides. Reports of the use of thiosemicarbazone complexes for this reaction have appeared only relatively recently. Representative metal complexes are shown in Figure 5; see also Figures 1, 3 and 4.

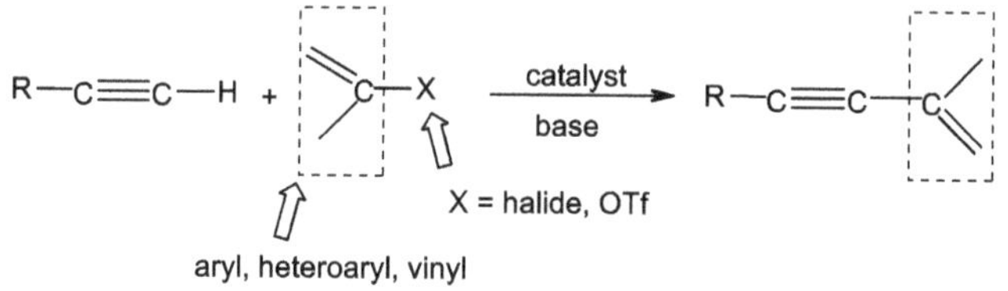

Scheme 24. Sonogashira reaction.

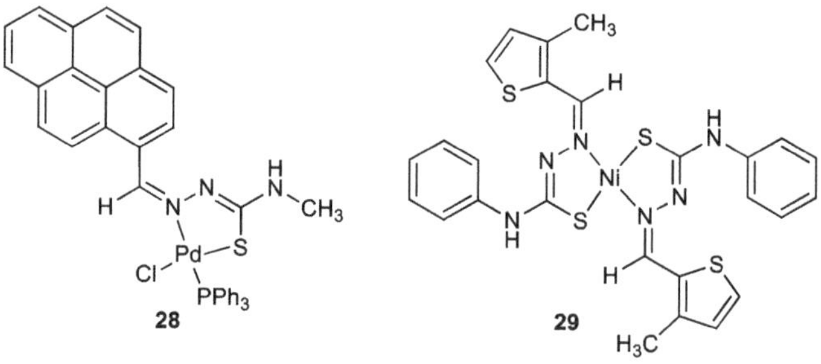

Figure 5. Representative metal complexes of thiosemicarbazones as catalysts for the Sonogashira reaction.

Reports of the use of thiosemicarbazone complexes for this reaction have appeared only relatively recently. The few studies that have been made concern Pd and Ni complexes and are often subsidiary to studies of other coupling reactions. Thus there have been no significant studies on the nature of the active species in these reactions or of other features which may confer advantages over previously reported complexes.

In 2011, Paul et al. in their study described above in connection with the use of thiosemicarbazone complexes of palladium in the Mizoroki–Heck or Suzuki–Miyaura couplings also examined their application to the Sonogashira reaction (see complexes **3** and **4** in Figure 1) [95]. Moderate to good conversions were obtained for the coupling of a limited number of aryl bromides with phenyl acetylene using either toluene-ethanol or PEG as solvent in the presence of Cu(I) and NaOH at 75–110 °C (Scheme 25). Catalyst loadings of 0.5 mol% were employed giving TONs of up to 200 and TOFs of up to 20 h^{-1}.

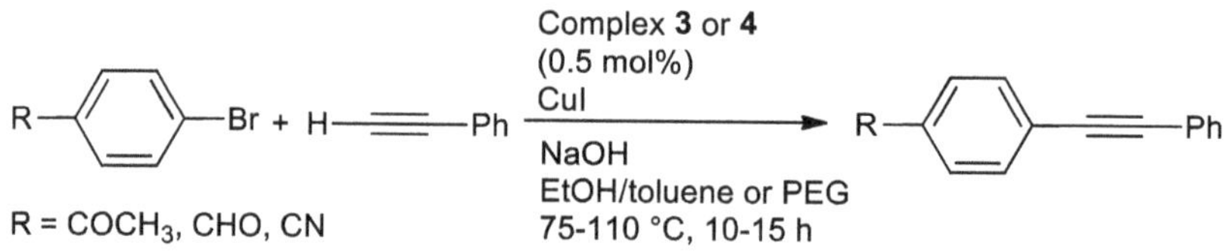

Scheme 25. Sonogashira reaction of aryl bromides with phenylacetylene catalysed by complex **3** or **4**.

In addition to the Suzuki–Miyaura reaction, Verma et al. also applied their carbohydrate derived thiosemicarbazone Pd complex **18**, shown in Figure 3, to the Sonogashira reaction between phenylacetylene and chlorobenzene, *p*-nitrobromobenzene or iodobenzene in triethylamine at 80 °C [111]. Moderate conversions of about 65% were obtained with 0.5 mol% catalyst loadings. In order to avoid the use of copper compounds in the Sonogashira reaction, a number of attempts have been made to develop complexes that are active under copper-free conditions. The first instance of a such a catalyst containing a thiosemicarbazone ligand was reported by Prabhu and Pal who synthesised a pyrenealdehyde thiosemicarbazonide palladium complex **28** (Figure 5) containing a Ph$_3$P supporting ligand [122]. Single crystal X-ray diffraction indicated bidentate *N,S*-coordination of the ligand. The complex is air stable and was shown to catalyse the Sonogashira reaction between phenylacetylene and a range of aryl chlorides and bromides at room temperature in DMF/Et$_3$N using a 0.5 or 1 mol% catalyst loading (Scheme 26). Moderate to very good conversions were obtained after 12 h (for the bromides) or 24 h (for the chlorides).

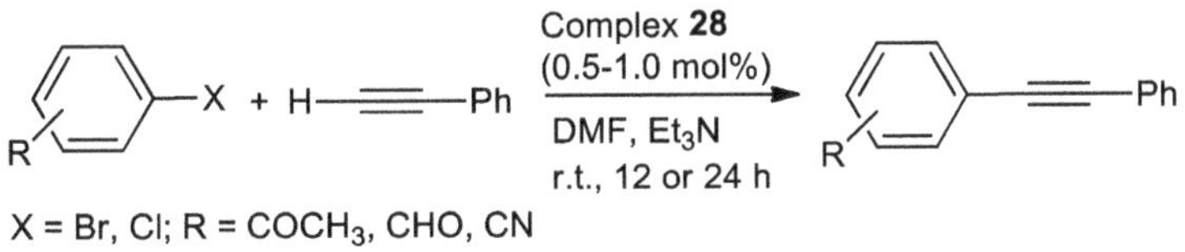

Scheme 26. Copper-free Sonogashira reaction of aryl halides with phenylacetylene catalysed by complex 28.

The octahedral nickel complexes 26 (Figure 4) prepared by Anitha et al. derived from 9,10-phenanthrenequinone thiosemicarbazone, 9,10-phenanthrenequinone N-methylthiosemi carbazone and 9,10-phenanthrenequinone N-phenylthiosemicarbazone described briefly above in connection with the Suzuki–Miyaura reaction were also examined for their activity in the Sonogashira reaction of phenyl acetylene with aryl halides [119]. Using catalyst loadings of 0.5 mol%, they were found to give good to very good conversions after 4 h in MeOH and in the presence of Cu(I) and pyridine (Scheme 27). Heteroaromatic chlorides also entered into the reaction as did ortho-substituted aromatics, albeit in lower yields. The authors concluded that steric effects in the ligands play a more important role than electronic effects in the catalytic activity of the complexes. Very good conversions were observed by Prabhu and Ramesh with a square-planar nickel complex NiL$_2$ (29) (Figure 5) where ligand L is derived from the reaction of 4-phenyl-3-thiosemicarbazide with 3-methyl-thiophene-2-carboxaldehyde [123]. The structure was confirmed by X-ray diffraction studies. Very promising results were obtained in the reaction of a range of aryl bromides and iodides with phenyl acetylene in the presence of the nickel complex together with Cu(I) in Et$_3$N at 80 °C (Scheme 28). Very good yields of the coupled products (79–99%) with good TONs were obtained after 2 h in the case of iodides (TONs of up to 1980 and TOFs of up to 990 h^{-1}) or 8 h in the case of the bromides (TONs of up to 990 and TOFs of up to 124 h^{-1}). Aryl halides with ortho-substitution also coupled readily.

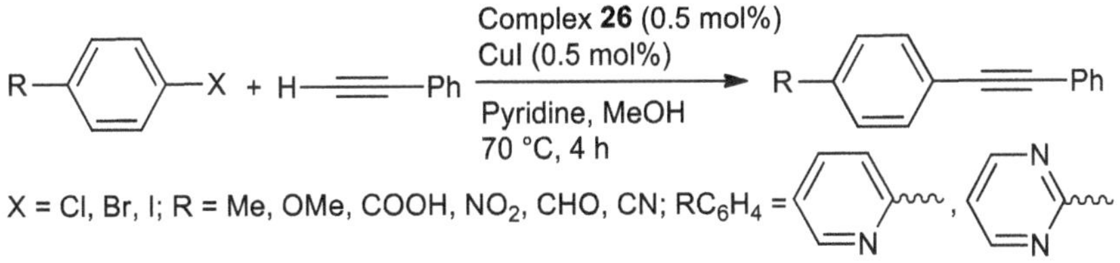

Scheme 27. Sonogashira reaction of aryl halides with phenylacetylene catalysed by Ni complexes 26.

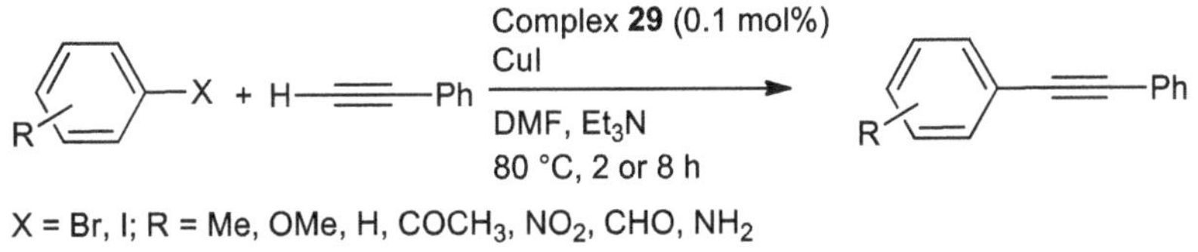

Scheme 28. Sonogashira reaction of aryl halides with phenylacetylene catalysed by Ni complex 29.

It is also appropriate to mention here a reaction related to the Sonogashira reaction, which involves coupling between aryl boronic acids with alkynes or alkynyl carboxylic acids (Scheme 29) reported by Lu et al. [124] using tridentate salicylaldiminato-thiosemicarbazone palladium catalysts 2 (Figure 1), which had previously been shown to catalyze the Mizoroki–Heck coupling [94]. The best yield was obtained by complex **2a**. Using mild conditions (CH$_2$Cl$_2$, KOAc, Ag$_2$O, 24 h, under argon) and a 2 mol% catalyst loading very good yields of coupled products were obtained except where steric hindrance was present (in the case of 1-naphthyl boronic acid and ortho-substituted aryl boronic acids) as well as for 2-pyridyl boronic acid. When carboxylic acids are used, decarboxylation occurs before coupling.

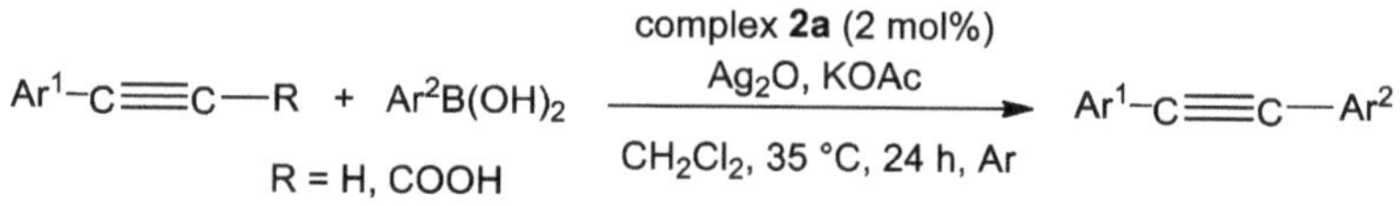

$$Ar^1-C\equiv C-R + Ar^2B(OH)_2 \xrightarrow[CH_2Cl_2,\ 35\ °C,\ 24\ h,\ Ar]{\text{complex 2a (2 mol\%)}\atop Ag_2O,\ KOAc} Ar^1-C\equiv C-Ar^2$$

R = H, COOH

Scheme 29. Alkynylation coupling reaction between alkynes or alkynyl carboxylic acids and arylboronic acids.

Additionally related to coupling reactions involving alkynes is the A3 coupling reaction, which is particularly useful in asymmetric synthesis [125]. This is a three-component reaction with an aldehyde, an amine and a terminal alkyne as the substrates. The reaction has been shown to be catalysed by a number of transition metal systems, including the thiosemicarbazone complex [Pd(PPh$_3$)L] **30** where L is a dianionic tridentate O,N,S-coordinating ligand derived from pyridoxal thiosemicarbazone or pyridoxal N-methylthiosemicarbazone as reported by Manikandan et al. (Scheme 30) [126]. In this case an ionic liquid, [emim]BF$_4$ (emim = 1-ethyl-3-methylimidazolium) was used as the reaction medium. After optimisation runs, a number of substrates were subjected to the reaction at 80 °C, 8 h reaction time with a 1 mol% catalyst loading. Phenyl acetylene was used as the terminal alkyne together with a range of aromatic or heteroaromatic aldehydes, formaldehyde or cyclohexyl carboxaldehyde and piperidine, morpholine, pyrrolidine or diethylamine as the amine component. In all cases, very good yields of the coupled products were obtained. Importantly, the catalyst could be recovered readily and retained its activity for at least five further cycles.

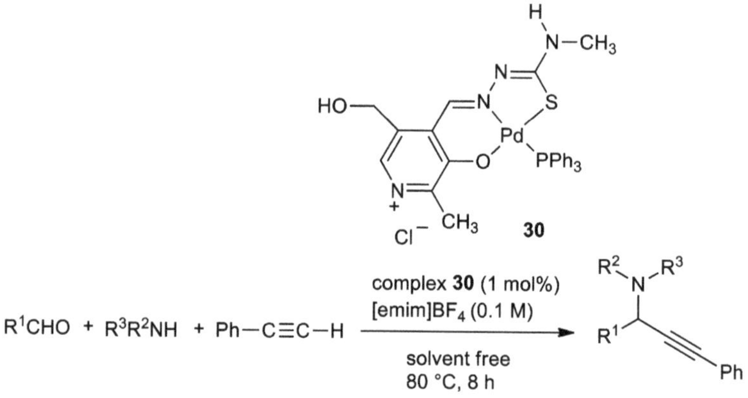

Scheme 30. A3 coupling reaction for the synthesis of propargylamines.

Sonogashira-type reactions catalysed by thiosemicarbazone complexes are summarised in Table 3.

Table 3. Sonogashira reactions catalysed by thiosemicarbazone complexes: representative conditions and yields [1].

Metal	T (°C)	Solvent	Time (h)	Ligand [2]	Base	Catalyst (mol%)	Yield (%)	Ref.
Pd	75–110	EtOH/toluene or PEG	10–15	N,S	NaOH	0.5	68–99	[95]
Pd	80	Et$_3$N	8	N,S	Et$_3$N	0.5	65	[111]
Pd	rt	DMF	12	N,S	Et$_3$N	0.5	67–99	[122]
Ni	70	MeOH	4	O,N,S	pyridine	0.5	55–85	[119]
Ni	80	DMF	8	N,S	Et$_3$N	0.1	79–99	[123]
Pd	35	CH$_2$Cl$_2$	24	O,N,S	KOAc	2.0	30–99	[124] [3]

[1] conditions refer to reactions involving aryl bromides and phenylacetylene. [2] ligand donor atoms. [3] reaction between arylboronic acids and phenylacetylene.

2.4. Kumada–Tamao–Corriu Reaction

The use of organometallic reagents to form carbon–carbon bonds is a standard procedure in organic synthesis but there are still many instances where the simple stoichiometric reaction is unsuccessful for one or more reasons. A number of transition metal catalysts have been developed for specific cases such as the Negishi coupling of organozinc reagents with aryl or alkenyl halides [127], or the related Kumada–Tamao–Corriu reaction involving the analogous coupling with Grignard reagents (Scheme 31) [8]. There are a number examples of the latter involving thiosemicarbazone complexes although the majority of the reports describe only one instance of a coupling of an aryl bromide and aryl magnesium bromide and thus do not permit a good assessment of wider applicability (Figure 6). Thus, there are accounts of ruthenium complexes derived from thiosemicarbazones. The mixed ligand complexes of Ru(II) **31**, [RuCO(EPh$_3$)$_2$L] and [RuCO(PPh$_3$)(py)L] (where E = P or As and L is a dibasic tridentate ligand derived from the condensation of ethylacetoacetate or methylacetoacetate and thiosemicarbazide) catalysed the coupling of PhMgBr and PhBr as reported by Thilagavathi et al. [64]. Using a 200:1 substrate to catalyst ratio, rather low conversions were reported. Analogous complexes derived from chalcone thiosemicarbazone gave similar results [128]. Somewhat better yields were reported by Raja et al. for the complexes **32** [RuCO(EPh$_3$)L] and [RuCO(py)L], where L is a tetracoordinated dianionic ligand derived from the reaction of 2-hydroxyaryl aldehyde, thiosemicarbazide and furfuraldehyde [62]. The coupling of PhMgBr with p-bromoanisole catalysed by a Ru(III) complexes **33** containing a monoanionic 2-acetylpyridine thiosemicarbazone ligand has also been reported by Manikandan et al. [81]. A 300:1 substrate to catalyst ratio was used and conversions of 28–48% were obtained.

Priyarega et al. reported nickel thiosemicarbazone complexes **34** with Ph$_3$P supporting ligand that catalyse the formation of biphenyl in good yield [129]. From the experimental data, it is stated that a large amount (0.05 mol) of complex is used for 0.01 mol of PhBr but presumably this is a typographical error with the correct amount of complex to be probably 0.05 mmol. Güveli et al. have prepared a series of thiosemicarbazone complexes **35** derived from o-hydroxyacetophenone in which either O,N,S- or O,N,N-tridentate coordination is observed [130]. In addition to structural and computational studies, the authors also examined the coupling of PhMgBr with PhBr in the presence of these compounds. The ONN-complexes gave higher yields compared to the ONS-complexes and this was ascribed to the larger size of the S atom and also to the higher charge on the metal. A range of aryl halides were employed by Anitha et al. in their study of Ni(II) complexes containing O,N,S-tricoordinating thiosemicarbazone ligands, which have also been described above as catalysts for the Suzuki–Miyaura and Sonogashira reactions (see complexes **26** in Figure 4) [119]. Moderate to excellent yields of biaryls were obtained under mild conditions (Et$_2$O, 4 h) and with catalyst loadings of 0.2 mol% (Scheme 32). TONs of up to 93 and TOFs of up to 2 h^{-1} were recorded. Reactions for aryl halides with electron withdrawing substituents were found to give slightly higher yields than those with electron-donating substituents, while ortho-substituted aryls gave lower yields. Overall, their catalytic efficiency was found to compare favorably with previously reported catalysts.

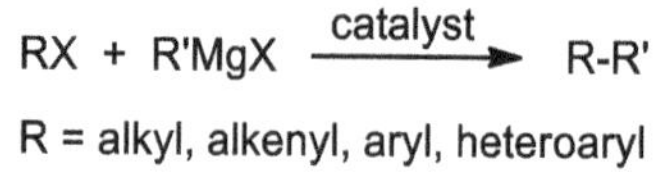
RX + R'MgX
catalyst
R-R'
R = alkyl, alkenyl, aryl, heteroaryl

Scheme 31. Kumada–Tamao–Corriu reaction.

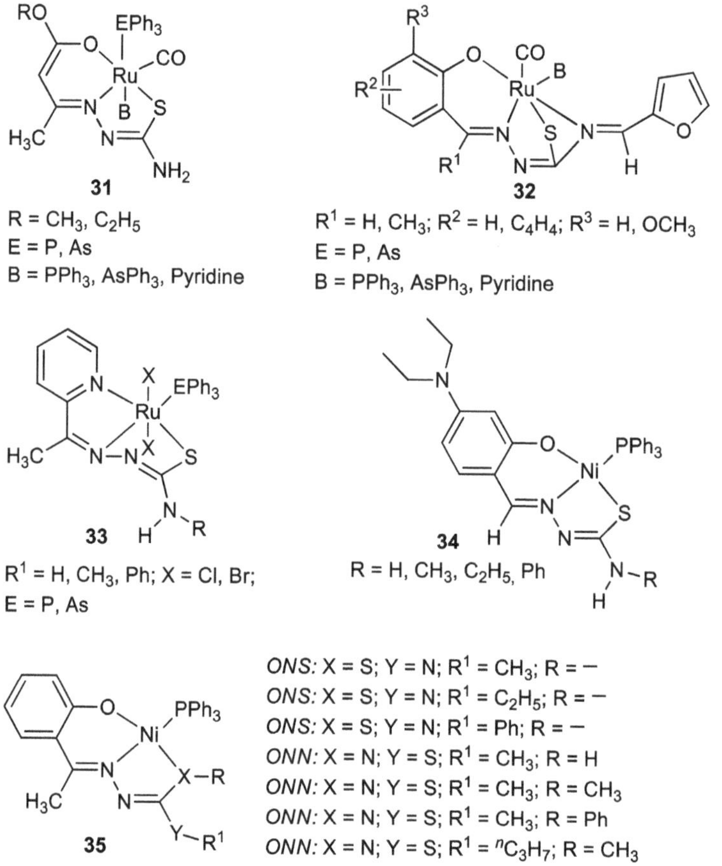
31
R = CH3, C2H5
E = P, As
B = PPh3, AsPh3, Pyridine
32
R1 = H, CH3; R2 = H, C4H4; R3 = H, OCH3
E = P, As
B = PPh3, AsPh3, Pyridine
33
R1 = H, CH3, Ph; X = Cl, Br;
E = P, As
34
R = H, CH3, C2H5, Ph
35
ONS: X = S; Y = N; R1 = CH3; R = −
ONS: X = S; Y = N; R1 = C2H5; R = −
ONS: X = S; Y = N; R1 = Ph; R = −
ONN: X = N; Y = S; R1 = CH3; R = H
ONN: X = N; Y = S; R1 = CH3; R = CH3
ONN: X = N; Y = S; R1 = CH3; R = Ph
ONN: X = N; Y = S; R1 = nC3H7; R = CH3

Figure 6. Representative metal complexes of thiosemicarbazones as catalysts for the Kumada–Tamao–Corriu reaction.

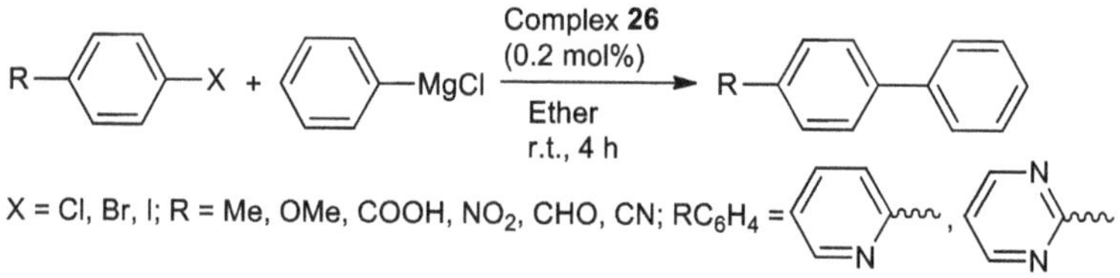
Complex 26
(0.2 mol%)
Ether
r.t., 4 h
X = Cl, Br, I; R = Me, OMe, COOH, NO2, CHO, CN; RC6H4 =

Scheme 32. Kumada–Tamao–Corriu reaction of aryl halides with phenylmagnesium chloride catalysed by Ni complexes **26**.

3. Carbon–Heteroatom Coupling Reactions

Although the majority of the work on metal complex catalysed coupling reactions concerns the formation of carbon–carbon bonds, carbon–heteroatom coupling reactions have also been widely studied. These predominantly concern the formation of carbon–nitrogen bonds for systems where non-catalysed coupling is not possible or very difficult. Reactions in this category include the Pd-catalysed arylation of amines (the Buchwald–Hartwig coupling shown in Scheme 33) [131,132], the Pd-catalysed formation of C–N, C–O or C–S bonds using aryl boronic acids and suitable heteroatom derivatives (the Chan-Lam coupling shown in Scheme 34) [133,134], and also the metal-catalysed Ullmann reaction (Scheme 35) [135,136]. Representative metal complexes catalyzed carbon–heteroatom coupling reactions are shown in Figure 7; see also in Figures 2 and 4.

A number of thiosemicarbazone complexes of palladium have been screened as potential catalysts for C–N coupling reactions. Many of these have also been investigated as catalysts for C–C couplings and have therefore been described above in the relevant sections. Thus, the mixed-ligand benzaldehyde thiosemicarbazone Pd-complexes **10**, **11**, **12**, and **13a** (Figure 2) prepared by Dutta et al., described previously [106,107], are also active catalysts for the Buchwald–Hartwig arylation of primary and secondary amines (Scheme 36). The addition of the hindered XPhos ligand (2-dicyclohexylphosphino-2′,4′,6′-triisopropylbiphenyl) was required for good activity and it was presumed that it was necessary to stabilize the active species, the other ligands having been displaced. Although somewhat higher catalyst loadings were found to be necessary for the C–N couplings than were used in the Suzuki–Miyaura reaction, catalytic efficiency was comparable to other palladium complexes under similar conditions. Very good results were also obtained under slightly milder conditions when the same group employed the cyclopalladated complexes **13b**, **14**, and **15** (Figure 2) for the coupling of selected aryl halides with aniline, providing TONs of 10,000 and TOFs of up to 833 h^{-1} (Scheme 37) [108]. A much broader range of aryl halides were used by Prabhu and Ramesh in their study of the catalytic activity of the complex **36** (Figure 7), [PdBr(PPh$_3$)L], where L is a bidentate chelating monoanionic ligand derived from 1-naphthaldehyde thiosemicarbazone [137]. Good to excellent results were obtained under relatively mild conditions (2-BuOH as solvent, K_2CO_3 as base, 100 °C, N_2 atmosphere, 24 h) using a 500:1 of substrate to complex molar ratio for the coupling of aromatic and heteroaromatic bromides with cyclic secondary amines (Scheme 38). Dibromides could also be successfully coupled with the same secondary amines under similar conditions. The coupling of aryl chlorides took place using slightly longer reaction times (30 h) and with somewhat lower conversions. It was found that the catalyst could be used twice without any detectable loss of activity but that gradual loss of activity was observed for subsequent cycles. Apart from the aforementioned monophosphine complexes, a recent study reports the synthesis of a cationic Pd thiosemicarbazone complex **23** (R = OCH_3; see in Figure 4) containing a diphosphine which, apart from catalysing the Suzuki–Miyaura reaction (see above) was also found to be effective in the Buchwald–Hartwig arylation [117]. Aryl bromides and iodides gave good conversions in the reaction with primary or secondary amines in either dioxane at 100 °C or PEG at 150 °C with low catalyst loadings (0.01 mol%) in the presence of NaOBu^t. Aryl chlorides, however, generally gave poor yields.

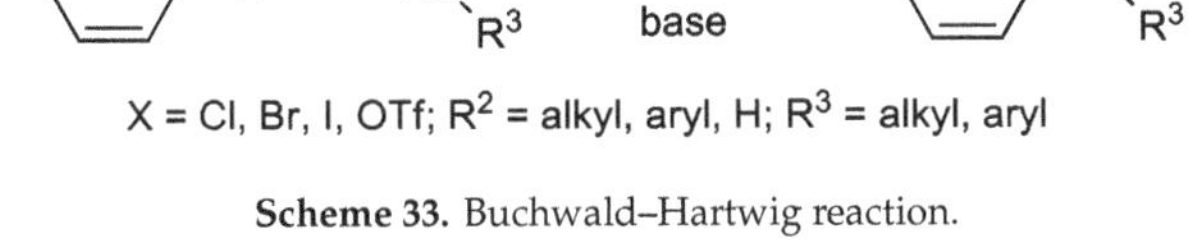

X = Cl, Br, I, OTf; R^2 = alkyl, aryl, H; R^3 = alkyl, aryl

Scheme 33. Buchwald–Hartwig reaction.

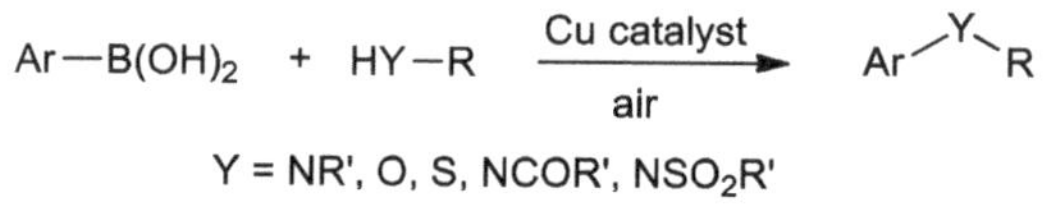

$$\text{Ar-B(OH)}_2 \ + \ \text{HY-R} \ \xrightarrow[\text{air}]{\text{Cu catalyst}} \ \text{Ar-Y-R}$$

Y = NR', O, S, NCOR', NSO$_2$R'

Scheme 34. Chan–Lam coupling.

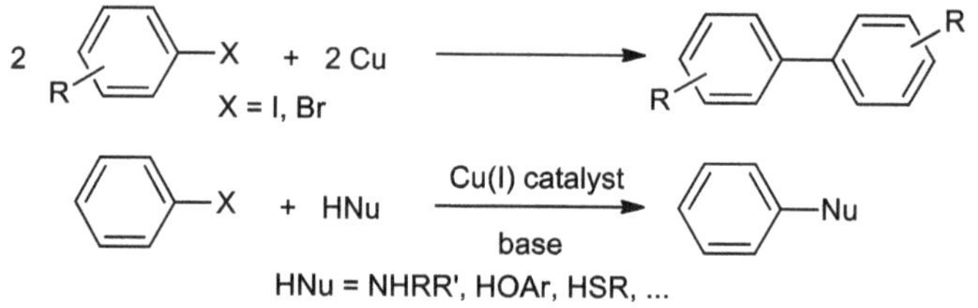

$$2 \ \text{R-C}_6\text{H}_4\text{-X} \ + \ 2 \ \text{Cu} \ \longrightarrow \ \text{R-C}_6\text{H}_4\text{-C}_6\text{H}_4\text{-R}$$

X = I, Br

$$\text{C}_6\text{H}_5\text{-X} \ + \ \text{HNu} \ \xrightarrow[\text{base}]{\text{Cu(I) catalyst}} \ \text{C}_6\text{H}_5\text{-Nu}$$

HNu = NHRR', HOAr, HSR, ...

Scheme 35. Classic Ullmann reaction (top) and Ullmann-type reaction (bottom).

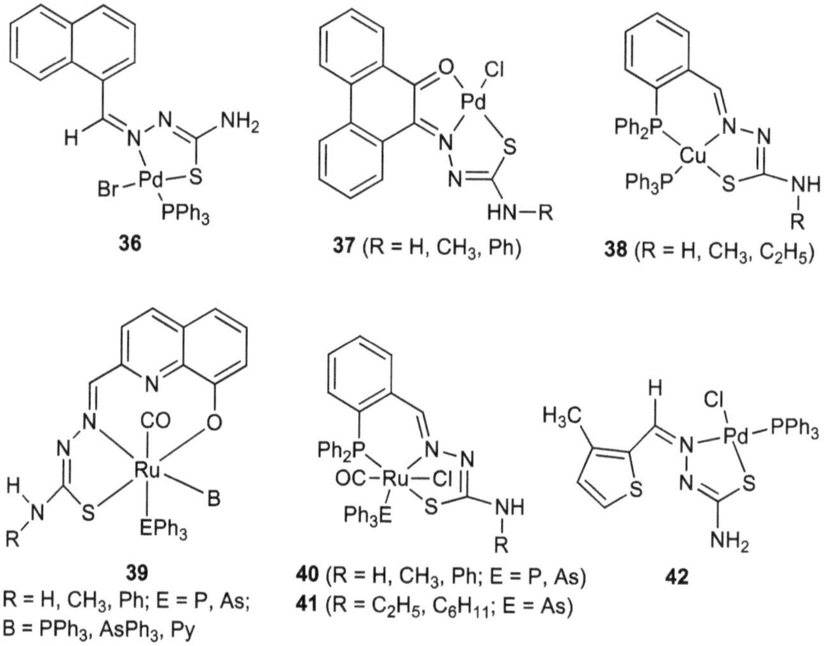

36

37 (R = H, CH$_3$, Ph)

38 (R = H, CH$_3$, C$_2$H$_5$)

39
R = H, CH$_3$, Ph; E = P, As;
B = PPh$_3$, AsPh$_3$, Py

40 (R = H, CH$_3$, Ph; E = P, As)
41 (R = C$_2$H$_5$, C$_6$H$_{11}$; E = As)

42

Figure 7. Representative metal complexes of thiosemicarbazones as catalysts for carbon–heteroatom coupling reactions.

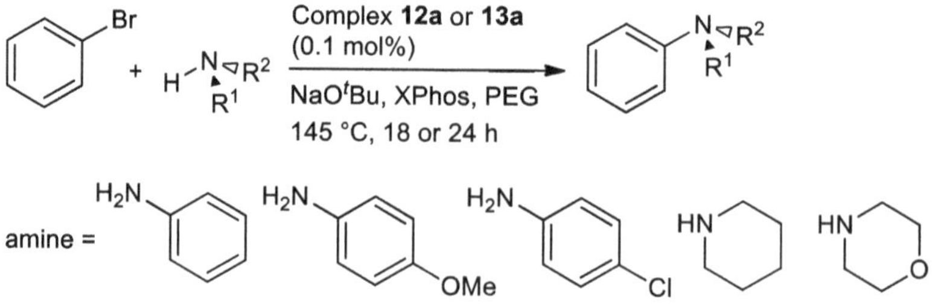

Scheme 36. Buchwald–Hartwig arylation of primary and secondary amines catalysed by complexes **12a** or **13a**.

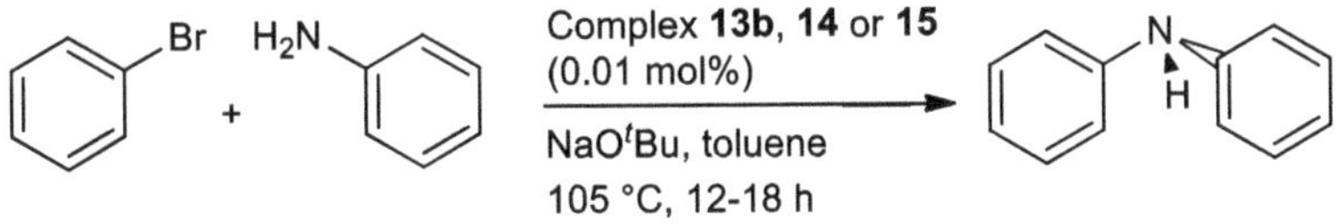

Scheme 37. Buchwald–Hartwig reaction of phenyl bromide with aniline catalysed by complexes 13b, 14, or 15.

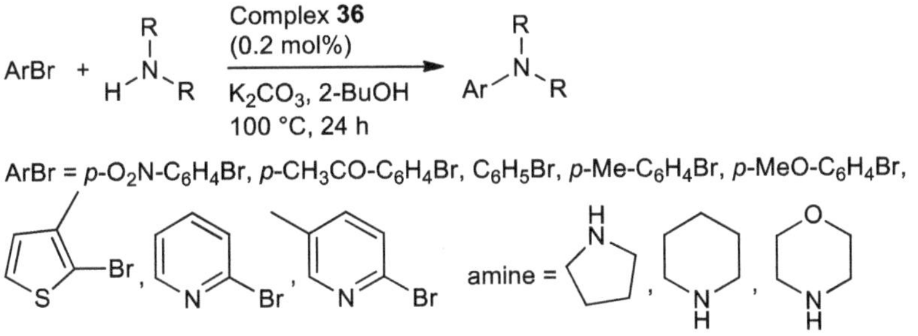

ArBr = p-O$_2$N-C$_6$H$_4$Br, p-CH$_3$CO-C$_6$H$_4$Br, C$_6$H$_5$Br, p-Me-C$_6$H$_4$Br, p-MeO-C$_6$H$_4$Br,

Scheme 38. Buchwald–Hartwig reaction of aryl- and heteroaryl bromides with cyclic secondary amines catalysed by complex 36.

N-arylation of heterocycles can been achieved by using a palladium catalyst derived from 9,10-phenanthrenequinone thiosemicarbazones 37 (Figure 7) as well as the corresponding semicarbazone as reported by Anitha et al. [138]. The ligand in these complexes is tridentate monoanionic and the most efficient one in initial screening experiments was that derived from phenanthrenequinone N-methylthiosemicarbazone. A range of aromatic and heteroaromatic chlorides, bromides and iodides were employed for coupling with imidazole in DMSO at 110 °C in the presence of KOH and 0.75 mol% of the complex. Moderate to very good yields were obtained, and it was shown that the reaction had the potential to be extended to other related heterocycles. Due to the fact that the best solvent for the reaction was DMSO, the authors favour a Pd(II)/Pd(IV) mechanistic oxidative addition pathway over a Pd(0)/Pd(II) one.

Copper complexes are also known to catalyse C–N coupling reactions. These are often Cu(I) complexes but the first instance of a thiosemicarbazone copper complex catalysed C–N coupling involves the Cu(II) oxidation state as reported by Shan et al. [139]. The use of copper compounds in this oxidation state offers some advantages since they are generally more convenient to handle. The catalytic procedure involved in situ complex formation from CuCl$_2$ (10 mol%) and excess ligand, 3-methoxy, 4-hydroxybenzaldehyde thiosemicarbazone in the presence of K$_2$CO$_3$ together with the appropriate substrates in DMF at 110 °C. Under these conditions, moderate to good yields of coupled products were obtained from the reaction of imidazole or benzimidazole with aryl bromides or iodides. The use of Cu(I) instead of CuCl$_2$ gave inferior results under the same conditions and the authors speculate that the reaction proceeds via a Cu(II)/Cu(IV) oxidative-addition pathway which is favoured by the stabilisation of the copper intermediate by the electron-rich ligand and the consequent decrease in the oxidation potential. Another instance of Cu(II) catalysed N-arylation was also recently reported by Gogoi et al. [140]. This involves a Chan-Lam coupling of a number of aryl or heteroaryl boronic acids with aniline or with N-containing heterocycles. Here again the complex was formed in situ, in this case from Cu(OAc)$_2$ and 2,5-dimethoxy benzaldehyde-4-phenylthiosemicarbazide; use of the preformed complex gave inferior results. Moderate to very good yields were obtained under mild conditions (room temperature, aqueous DMF as solvent, Et$_3$N as base, 10 mol% catalyst loading) which compare very favorably with previously reported results for similar systems.

N-alkylation of amines by means of alkyl alcohols can be catalysed by Cu(I) complexes and this has been demonstrated for complexes 38 (Figure 7) containing 2-(2-(diphenylphosphino) benzylidene)

thiosemicarbazone ligands by Ramachandran et al. (Scheme 39) [141]. In these complexes, the ligand is bound to the metal through P, N and S. A range of substituted benzyl alcohols were employed together with a number substituted aminobenzothiazoles as well as 1-amino-diphenylthiazole, benzimidazole and 2,6-diaminopyridine. n-Butanol and n-hexanol were also successfully used as the alkylating agent. The conditions used involved a catalyst loading of 0.1 mol%, KOH as base, toluene as solvent and heating to 100 °C for 12 h. Good to excellent yields, TONs of up to 990 and TOFs of up to 83 h^{-1} were obtained, in particular for complex in which R is the CH$_3$ group.

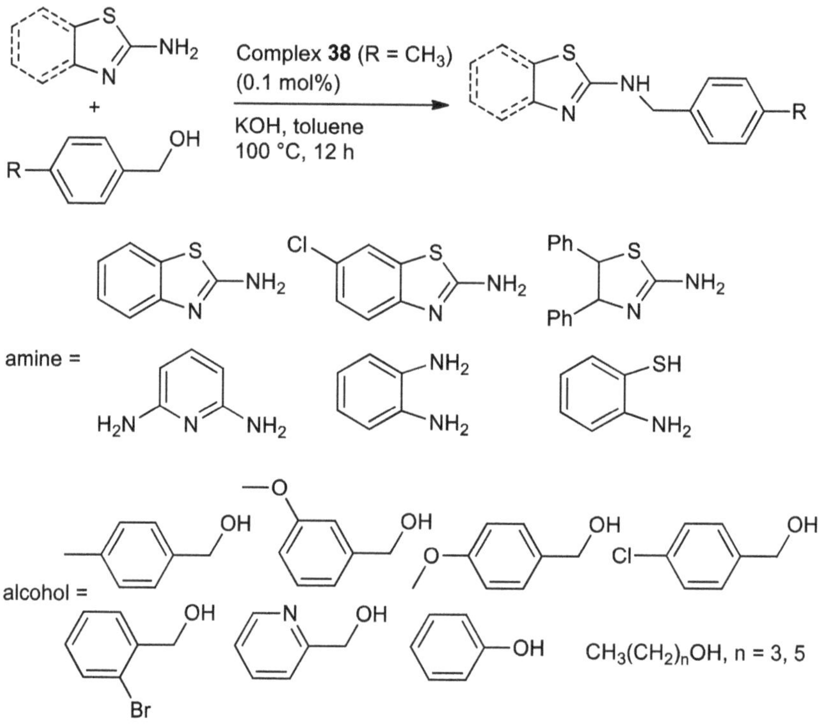

Scheme 39. N-alkylation of amines with various alcohols catalysed by Cu complex **38** (R = CH$_3$).

Several reports have also appeared from the group of Viswanathamurthi and co-workers concerning C–N coupling reactions catalysed by ruthenium complexes [74,142–145]. The initial report briefly describes the benzylation of aniline catalysed by ruthenium hydroxyquinoline–thiosemicarbazone complexes **39** (Figure 7) at 100 °C under nitrogen in the presence of KOBu^t. Good conversions were obtained using a 1000:1 substrate to catalyst molar ratio [74]. Subsequent reports from this group describe XRD structurally characterised complexes containing tridentate P,N,S-chelating thiosemicarbazone derivatives of 2-diphenylphosphino benzaldehyde as ligands in which the effect of terminal N-substitution is examined. Thus, the complexes [RuCl(CO)(EPh$_3$)L] (**40**) (Figure 7), where E = P, As and L = 2-(2-(diphenylphosphino) benzylidene)-N-R-thiosemicarbazone (R = H, CH$_3$ or Ph) were prepared and their catalytic activity studied for the N-alkylation of heteroaromatic amines by alcohols [142]. Optimisation experiments indicated that the complex containing the 2-(2-(diphenylphosphino)benzylidene)-N-methylthiosemicarbazone was found to give the best results, with a 0.5 mol% catalyst loading in the presence of KOH in toluene at 100 °C for 12 h (Scheme 40). Good to very good yields were obtained for the alkylation using p-C$_6$H$_4$CH$_2$OH or ferrocenylCH$_2$OH and primary amines such as aniline, 2-aminopyridine 2-aminobenzothiazole while dialkylation of 2,6-diaminopyridine could also be affected. In addition to straightforward alkylations, when 2-nitropyridine was employed as substrate initial reduction to the primary amine and subsequent alkylation could be achieved, while for aminobenzene ortho-substituted with NH_2, OH or SH, the alkylation reaction with primary alcohols gave good yields of 2-substituted heterocyclic products, viz. benzazoles, benzoxazoles, or benzothiazoles. In a subsequent report, comparable catalytic activity was demonstrated for complexes of the type [RuCl(CO)(AsPh$_3$)(L)] (**41**) where L = 2-(2-(diphenylphosphino)benzylidene)-*N*-ethyl-thiosemi carbazone or 2-(2-(diphenylphosphino) benzylidene)-*N*-cyclohexyl-thiosemicarbazone (Figure 7) [143]. The authors propose that the mechanism for the reaction is via a so-called "borrowed hydrogen" pathway, whereby the alcohol is catalytically dehydrogenated to the corresponding aldehyde, which then condenses with the amine to give an intermediate imine, which is subsequently hydrogenated by the catalyst. A further series of ruthenium complexes, bearing the 2-(2-(diphenylphosphino) benzylidene)-*N*-ethyl-thiosemicarbazone ligand, L, were prepared, namely [RuCl(CO)(PPh$_3$)L], [RuH(CO)(PPh$_3$)$_2$L], [RuCl(PPh$_3$)$_2$L], [RuCl(dmso)$_2$L], and [RuL$_2$] [144]. Of these complexes, the last one showed little or no catalytic activity while the first three complexes showed the best results and were studied in more detail. Using conditions comparable with those in the previous studies, and with similar substrates, the complexes were found to give good to excellent conversions. *N*-Alkylation of sulfonamides was also very successful. We may also mention here that the complex [RuL$_2$], where L is the monoanionic ligand derived from 2-(2-(diphenylphosphino)-benzylidene)-*N*-phenylthiosemicarbazone, has also been screened for its catalytic activity in the *N*-alkylation of primary amines [145]. In this investigation, the complex proved to be inferior to other semicarbazone complexes that were examined and was not subjected to further detailed study.

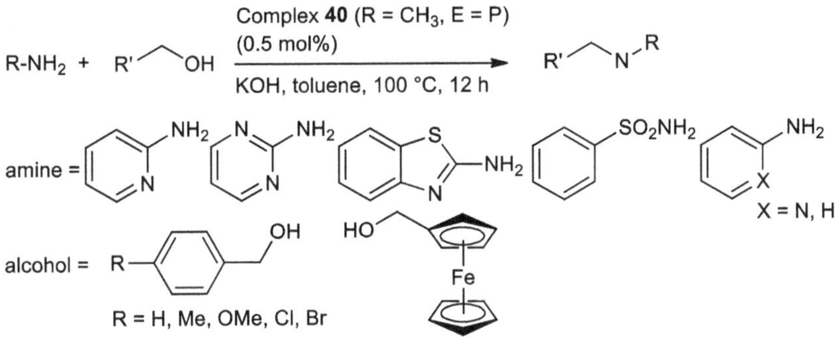

Scheme 40. *N*-alkylation of (hetero)aromatic amine/amides with alcohols catalysed by Ru complex **40** (R = CH$_3$, E = P).

Apart from the above C–N coupling reactions, there has also been one report by Suganthy et al. of C–O coupling catalysed by a thiosemicarbazone complex [146]. Thus the coupling of *p*-cresol with a number of aryl halides containing electron-withdrawing or electron-donating groups could be effected, with moderate to very good yields, in DMF at 80 °C after 12 h in an inert atmosphere using a 1 mol% loading of the catalyst [PdCl(PPh$_3$)L] (**42**), where L is the bidentate *N*,*S*-chelating ligand derived from the deprotonation of 3-methyl-thiophene-2-carboxaldehyde thiosemicarbazone (Figure 7).

Published results of the use of thiosemicarbazone complexes as described above for C-heteroatom coupling reactions are summarised in Table 4.

Table 4. Carbon–heteroatom coupling reactions catalysed by thiosemicarbazone complexes: representative conditions and yields.

Metal	Substrates	T (°C)	Solvent	Time (h)	Ligand [1]	Base	Catalyst (mol%)	Yield (%)	Ref.
Pd	ArBr + 2^{ary} amine	145	PEG	24	N,S	NaOBu^t	1.0	50–62	[106]
Pd	ArBr + 1^{ary}/2^{ary} amine	145	PEG	24	N,S	NaOBu^t	0.1	100	[107]
Pd	ArBr + 1^{ary}/2^{ary} amine	145	PEG	18	C,N,S	NaOBu^t	0.1	100	[107]
Pd	ArBr + aniline	105	toluene	12–18	C,N,S	NaOBu^t	0.01	100	[108]
Pd	(het)ArBr + 2^{ary} amine	100	2-BuOH	24	N,S	K$_2$CO$_3$	0.2	77–99	[137]
Pd	(het)ArBr + N-heterocycle	110	DMSO	10	O,N,S	KOH	0.75	75–90	[138]
Cu	ArBr + N-heterocycle	110	DMF	24	O,N,S	K$_2$CO$_3$	10.0	42–56	[139]
Cu	(het)ArB(OH)$_2$ + aniline	r.t.	DMF/H$_2$O	14–18	N,S	Et$_3$N	10.0	74–95	[140]
Cu	(het)ArB(OH)$_2$ + N-heterocycle	r.t.	DMF/H$_2$O	18–24	N,S	Et$_3$N	10.0	70–94	[140]
Cu	RCH$_2$OH + 1^{ary} amine	100	toluene	12	P,N,S	KOH	0.1–0.2	89–99	[141]
Ru	RCH$_2$OH + aniline	100	none	6	O,N,S	KOBu^t	1.0	61–86	[74]
Ru	RCH$_2$OH + 1^{ary} amine	100	toluene	12–24	P,N,S	KOH	0.5–1.0	45–99	[142]
Ru	RCH$_2$OH + 1^{ary} amine	100	toluene	12	P,N,S	KOH	0.5	79–98	[143]
Ru	RCH$_2$OH + 1^{ary} amine	100	toluene	12	P,N,S	KOH	0.5	59–98	[144]
Ru	RCH$_2$OH + sulfonamide	120	toluene	12	P,N,S	KOH	0.5	21–99	[144]
Pd	ArBr/ArI + p-cresol	80	DMF	12	N,S	K$_2$CO$_3$	1.0	62–94	[146]

[1] ligand donor atoms.

4. Immobilised and Heterogeneous Catalysts

Recovery of catalysts after use is an important factor for consideration for all reactions which involve transition metal complexes. Not only it is important because of the often high cost of the catalysts themselves, but it is also important to minimise contamination of the products. This becomes particularly significant when scale-up of a reaction is being planned. For these reasons, a great deal of effort has been made to develop heterogeneous analogues of homogeneous catalytic reactions in which, the catalyst can be reclaimed by straightforward separation procedures although, attractive as it may seem, this is not always without its disadvantages [147], and special consideration must be given to the stability of the catalysts and to leaching phenomena [87]. In the case of thiosemicarbazone complexes, there have been a number of attempts to develop such systems for coupling reactions. By reduction of K$_2$PdCl$_4$ with hydrazine hydrate in the presence of pyridine-2-carbaldehyde thiosemicarbazone as stabilizer of the nanoparticles, Kostas, Kovala-Demertzi, and co-workers were able to prepare nanoparticles which were characterized by XRD and SEM [148]. They were active catalysts for the Suzuki–Miyaura reaction of phenyl boronic acid with aryl bromides (Scheme 41). Best results were obtained for p-bromonitrobenzene and p-bromobenzonitrile, which gave excellent conversions in DMF/H$_2$O at 100 °C with 0.1% w/w catalyst loading and with K$_2$CO$_3$ as base. Higher catalyst loadings (1% w/w) were required for good yields from the coupling reactions with bromobenzene and p-bromoanisole, as well as for reactions at room temperature, which only gave good yields with p-BrC$_6$H$_4$NO$_2$ and p-BrC$_6$H$_4$CN. These thiosemicarbazone-derivatized nanoparticles were found to be more efficient catalysts than the homogeneous catalyst Pd(PPh$_3$)$_4$ under identical reaction conditions. The catalyst could be recovered but was progressively less active in successive cycles.

Bakherad et al. have reported a polystyrene supported complex of palladium with a 1-phenyl-1,2-propanedione-2-oxime thiosemicarbazone ligand [149]. By attaching the ligand to the polystyrene and then reaction with PdCl$_2$(PhCN)$_2$ followed by reduction with hydrazine hydrate, a Pd(0) species was obtained which was evaluated for its catalytic activity in the acylation of terminal alkynes. Under optimized solvent-free conditions, namely using 1 mol% catalyst with Et$_3$N as base, excellent conversions (97–99%) were obtained after 30 min in air at room temperature for a range of aromatic acyl chlorides as well as cyclohexyl carboxylic acid chloride with phenylacetylene, pent-1-yne, hex-1-yne, and Me$_3$SiC≡CH (Scheme 42). The catalyst could be recovered by centrifugation and was reused several times with only a slight decrease in activity.

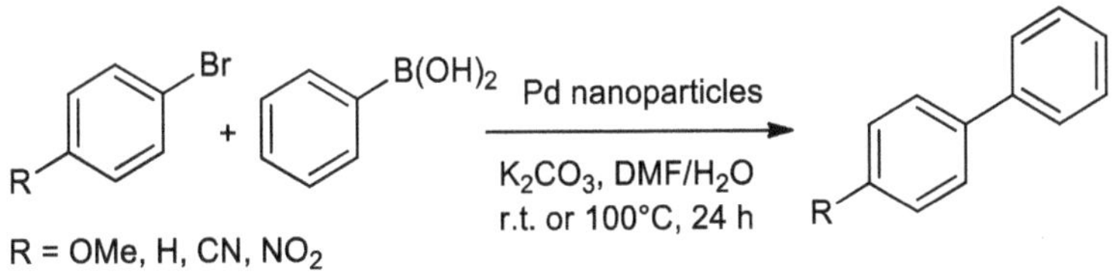

Scheme 41. Suzuki reaction of aryl bromides with phenylboronic acid catalysed by thiosemicarbazone-derivatised Pd nanoparticles.

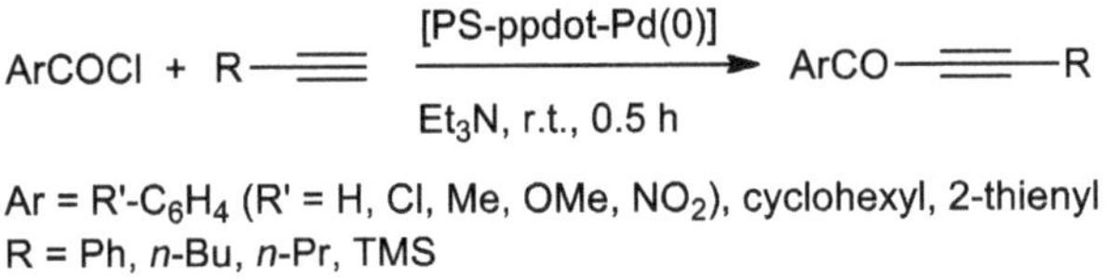

ArCOCl + R—≡ $\xrightarrow[\text{Et}_3\text{N, r.t., 0.5 h}]{\text{[PS-ppdot-Pd(0)]}}$ ArCO—≡—R

Ar = R'-C$_6$H$_4$ (R' = H, Cl, Me, OMe, NO$_2$), cyclohexyl, 2-thienyl
R = Ph, n-Bu, n-Pr, TMS

Scheme 42. Copper- and solvent-free Sonogashira reaction of acid chlorides with terminal alkynes catalysed by 1-phenyl-1,2-propanedione-2-oxime thiosemicarbazone-functionalized polystyrene resin-supported Pd(0) complex [PS-ppdot-Pd(0)].

Suzuki–Miyaura coupling of various aryl halides with alkenyl boronic acid has been achieved using a Pd(II) thiosemicarbazone complex tethered to a silica support [150]. Aryl halides were coupled with trans-2-phenylvinyl boronic acid in DMF/H$_2$O in the presence of K$_2$CO$_3$ and catalyst (25 mg per mmol of ArX), under microwave irradiation at 110 °C for 25 min. The reaction was selective for the formation of E-stilbenes and the catalyst was readily recovered by filtration. The catalyst could be used for at least six consecutive trials without loss of activity. Absence of palladium in the liquid phase after filtration suggested that no leaching of the catalyst had occurred during the reaction.

Veisi et al. have used multi-walled carbon nanotubes to which thiosemicarbazide has been grafted to form supported Cu(I) complexes that are able to catalyse the Ullmann coupling of indole, amines or imidazoles with aryl halides [151]. The coupling reactions were optimally carried out in DMF/Et$_3$N at 80 °C using a substrate:Cu ratio of 50/1 and reaction times ranging from 1–3 h for ArI, 3–6 h for ArBr and 12–24 h for ArCl (Scheme 43). Good to excellent yields (65–98%) were reported. The catalyst could be recovered by centrifugation and, in studies of the coupling of PhI with indole, was found to be reusable five times with marginal loss of activity. The filtrate from the reaction was found to be inactive, indicating that no leaching of active complex from the supported complexes had occurred.

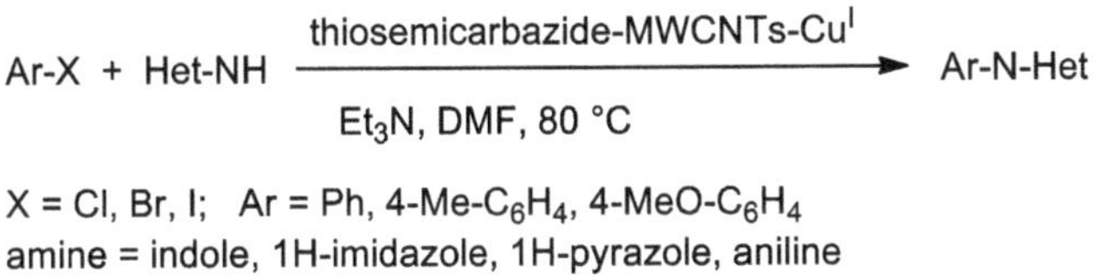

Ar-X + Het-NH $\xrightarrow[\text{Et}_3\text{N, DMF, 80 °C}]{\text{thiosemicarbazide-MWCNTs-Cu}^{\text{I}}}$ Ar-N-Het

X = Cl, Br, I; Ar = Ph, 4-Me-C$_6$H$_4$, 4-MeO-C$_6$H$_4$
amine = indole, 1H-imidazole, 1H-pyrazole, aniline

Scheme 43. Ulmann coupling of indole, amines, or imidazoles with aryl halides catalysed by thiosemicarbazide-multi-walled carbon nanotubes-Cu^{I} nanocatalyst.

Halloysite is a form of natural clay that can be modified by the attachment of a variety of functionalities. Sadjadi has reported the conjugation of tosylated cyclodextrin to thiosemicarbazide functionalized halloysite, which in turn was used to immobilize Pd nanoparticles [152]. This immobilized system was then examined for its activity in the Sonogashira and Mizoroki–Heck reactions. In the former case, a range of aryl halides were coupled with phenylacetylene or propargyl alcohol in water/EtOH at 60 °C, in the presence of K_2CO_3 using 6 mol% Pd catalyst loading (Scheme 44). The activities followed the usual trend with iodides requiring the shortest reaction times (1.5–3.5 h) and giving the best conversions (83–95%) and chlorides requiring the longest reaction times (ca. 5 h) and giving moderate conversions (ca. 50%). In an examination of the recyclability of the catalyst, it was recovered and reused thirteen times in a typical reaction. The first four runs showed comparable activity but subsequent runs showed a gradual reduction such that from an original 95% conversion the final run gave 69%. In order to obtain more insight into the nature of the catalyst, the authors examined the recycled material by SEM, TEM and FT-IR. They found that there were no major observable differences after four cycles but thereafter there were indications of morphological changes due to agglomeration although the basic structure of the material was maintained. This agglomeration together with a limited amount of leaching that was also detected was presumed to be the cause of the gradual decrease in activity. The immobilised system was also demonstrated to be an active catalyst for the Mizoroki–Heck coupling of iodobenzene with styrene but a systematic examination of the scope of the reaction was not performed. Halloysite has also been functionalized with (3-chloropropyl) trimethoxysilane and subsequently reacted with thiosemicarbazide and furfural and then with $Cu(OAc)_2$ to provide an immobilized copper species [153]. Using ultrasonic irradiation, this system was active in the A3 coupling reactions of aldehydes, phenyl acetylene, and amines for synthesis of corresponding propargylamines. Very good conversions were obtained at room temperature within 30 min. The catalyst was readily recovered and after four runs showed little loss in activity. No leaching was detected and the catalyst reclaimed after successive runs and examination by FTIR, TEM, XRD, SEM, and EDX indicated it to be essentially unchanged.

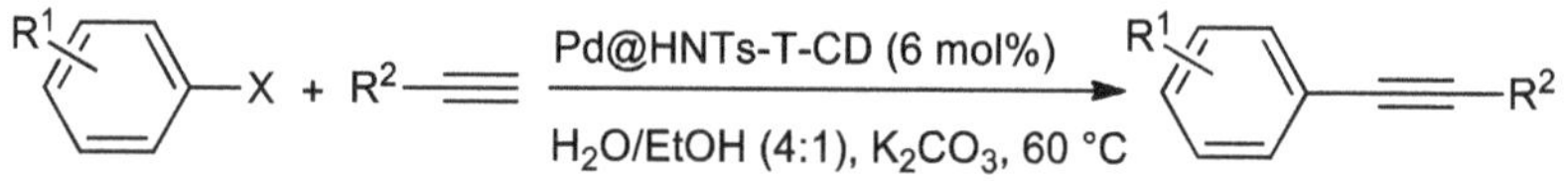

$X = Cl, Br, I; R = H, Me, OMe, OH, CH_3CO, NH_2, NO_2, CHO; R^1C_6H_4 = $ 1-naphthyl
$R^2 = Ph, HOCH_2$

Scheme 44. Sonogashira reaction of aryl halides with terminal alkynes catalysed by Pd nanoparticles immobilized on tosylated cyclodextrin-thiosemicarbazide-functionalized-halloysite nanotubes (Pd@HNTs-T-CD).

Finally, there have recently been reports on the use of thiosemicarbazide modified mesoporous silica materials in the immobilization of potential catalysts. Sadjadi et al. have prepared such a system, which consists of furfural thiosemicarbazone tethered through the terminal nitrogen to the mesoporous SBA-15 [154]. Reaction with copper acetate produced a system which, similar to that described above, was an active A3 coupling catalyst at ambient temperatures. Using a catalyst loading of 0.5 mol%, and under solvent-free conditions, very good yields of products, TONs of up to 190, and TOFs of up to 13 min^{-1} were obtained within 40 min or less for the coupling of a variety of aryl aldehydes, phenyl acetylene, and morpholine or piperidine (Scheme 45). The reusability of the catalyst was also confirmed for four cycles. A similar copper system derived from the mesoporous material SBA-16 was used for C–S arylation reactions as reported by Ghodsinia et al. [155]. Coupling of aryl halides with elemental sulfur or thiourea provide symmetrical diaryl sulfides in generally very good yields. A catalyst loading of 1.3 mol% was used under solvent-free conditions in the presence of KOH. The yields in the reaction increased according to the aryl halide, ArX used, in the order Cl < Br < I. A study of the reusability of the catalyst indicated only a gradual loss of activity after seven runs. An array of techniques were employed to investigate the catalyst and it was found that the structural integrity of the material was maintained after successive runs. In addition, no significant leaching was detected. It was suggested that the loss of activity that had been observed was due to the partial saturation during the reaction process of the mesochannels containing the catalytic active sites. A very recent paper by Ahmadi et al. describes a magnetic mesoporous silica-Fe_3O_4 nanocomposite functionalised with a Pd thiosemicarbazone complex [156]. The material was investigated for its catalytic activity in the Suzuki–Miyaura reaction. Optimal conditions were found to be DMF as solvent and a temperature of 120 °C. The preferred base was K_2CO_3 (1.2 mmol) and a 0.18 mol% (based on Pd) catalyst loading was used. A variety of aryl halides were examined in the reaction with phenyl boronic acid. With the exception of the hindered ortho-bromotoluene, all the halides used gave very good to excellent yields in short reaction times (60 min or less). After the catalytic run, the catalyst could be extracted using an external magnet. After washing and drying, the catalyst was reused and it was shown that it could be recycled for five times without significant decrease in the catalytic activity. Leaching was negligible and FT-IR and XRD indicated the structure of the catalyst to be unchanged after each cycle.

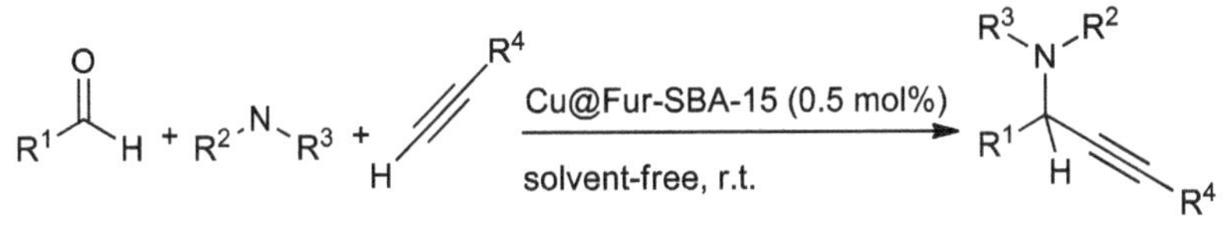

R^1 = C_6H_5, 4-ClC_6H_4, 4-HOC_6H_4, 2-HOC_6H_4, 4-O_2N-C_6H_4, 4-MeC_6H_4, 4-$MeOC_6H_4$
$R^2 R^3$ = -$(CH_2)_2$-O-$(CH_2)_2$-, C_5H_5; R^4 = C_6H_5

Scheme 45. A3 coupling reaction for the synthesis of propargylamines catalysed by Cu species immobilized on functionalized mesoporous SBA-15 with thiosemicarbazide and furfural (Cu@Fur-SBA-15).

5. Future Prospects

It is clear that thiosemicarbazone complexes are promising catalysts for a number of applications. Phosphane-free thiosemicarbazone complexes as well as the analogous complexes with additional P-ligands as catalysts for cross-coupling reactions have received much attention during the last fifteen years. The fact that the ligands are relatively readily accessible and that the complexes formed show good stability make them popular subjects for investigation. Up until now, there have been relatively few reports concerning the nature of the species formed during the reactions using these complexes. More research into aspects such as the formation of nanoparticles, aggregation and deaggregation phenomena, leaching effects, the role of the ligands with different metals etc. is needed for the development of systems with general applicability. Undoubtedly, there will be increasing attention paid to areas such as catalyst immobilisation and to complexes that are active under mild, preferably aerobic conditions. The potential for applying these complexes in solvent-free or aqueous systems is also clear. An additional area where developments may be expected is that of asymmetric catalysis. Up until now there appear to be no reports of thiosemicarbazone complexes having been investigated for this purpose but chiral thiosemicarbazone ligands have certainly been prepared and it remains to be seen whether their complexes can show the appropriate selectivity.

Author Contributions: I.D.K. and B.R.S. contributed equally. Conceptualization, I.D.K. and B.R.S.; writing—original draft preparation, I.D.K. and B.R.S.; writing—review and editing, I.D.K. and B.R.S. All authors have read and agreed to the published version of the manuscript.

Funding: This research received no external funding.

Conflicts of Interest: The authors declare no conflict of interest.

References

1. Biffis, A.; Centomo, P.; Del Zotto, A.; Zecca, M. Pd Metal Catalysts for cross-couplings and related reactions in the 21st century: A critical review. *Chem. Rev.* **2018**, *118*, 2249–2295. [CrossRef]
2. Beletskaya, I.P.; Alonso, F.; Tyurin, V. The Suzuki-Miyaura reaction after the Nobel prize. *Coord. Chem. Rev.* **2019**, *385*, 137–173. [CrossRef]

3. El-Maiss, J.; Mohy El Dine, T.; Lu, C.-S.; Karamé, I.; Kanj, A.; Polychronopoulou, K.; Shaya, J. Recent advances in metal-catalyzed alkyl–boron (C(sp^3)–C(sp^2)) Suzuki-Miyaura cross-couplings. *Catalysts* **2020**, *10*, 296. [CrossRef]
4. Beletskaya, I.P.; Averin, A.D. New trends in the cross-coupling and other catalytic reactions. *Pure Appl. Chem.* **2017**, *89*, 1413–1428. [CrossRef]
5. Kanwal, I.; Mujahid, A.; Rasool, N.; Rizwan, K.; Malik, A.; Ahmad, G.; Shah, S.A.A.; Rashid, U.; Nasir, N.M. Palladium and copper catalyzed Sonogashira cross coupling an excellent methodology for C-C bond formation over 17 years: A review. *Catalysts* **2020**, *10*, 443. [CrossRef]
6. Choi, J.; Fu, G.C. Transition metal–catalyzed alkyl-alkyl bond formation: Another dimension in cross-coupling chemistry. *Science* **2017**, *356*. [CrossRef] [PubMed]
7. Kumar, S. Recent advances in the Schiff bases and N-heterocyclic carbenes as ligands in the cross-coupling reactions: A comprehensive review. *J. Heterocycl. Chem.* **2019**, *56*, 1168–1230. [CrossRef]
8. Heravi, M.M.; Zadsirjan, V.; Hajiabbasi, P.; Hamidi, H. Advances in Kumada–Tamao–Corriu cross-coupling reaction: An update. *Mon. Chem. Chem. Mon.* **2019**, *150*, 535–591. [CrossRef]
9. Sain, S.; Jain, S.; Srivastava, M.; Vishwakarma, R.; Dwivedi, J. Application of palladium-catalyzed cross-coupling reactions in organic synthesis. *Curr. Org. Synth.* **2020**, *16*, 1105–1142. [CrossRef]
10. Mannepalli, L.K.; Gadipelly, C.; Deshmukh, G.; Likhar, P.; Pottabathula, S. Advances in C-C coupling reactions catalyzed by homogeneous phosphine free palladium catalysts. *Bull. Chem. Soc. Jpn.* **2020**, *93*, 355–372. [CrossRef]
11. Zimmer, M.; Schulte, G.; Luo, X.L.; Crabtree, R.H. Functional-modeling of Ni,Fe Hydrogenases: A nickel-complex in an N,O,S environment. *Angew. Chem. Int. Ed.* **1991**, *30*, 193–194. [CrossRef]
12. Vetter, A.H.; Berkessel, A. Nickel Complex Catalyzed Reduction of Imines. *Synthesis (Stuttg)* **1995**, *1995*, 419–422. [CrossRef]
13. Berkessel, A.; Hermann, G.; Rauch, O.-T.; Büchner, M.; Jacobi, A.; Huttner, G. Preparation and X-ray crystal structure of the first trimeric nickel thiosemicarbazone complex: The first example of oligomerization by both Ni–O–Ni and Ni–S–Ni bridging. *Chem. Ber.* **1996**, *129*, 1421–1423. [CrossRef]
14. Pelagatti, P.; Venturini, A.; Leporati, A.; Carcelli, M.; Costa, M.; Bacchi, A.; Pelizzi, G.; Pelizzi, C. Chemoselective homogeneous hydrogenation of phenylacetylene using thiosemicarbazone and thiobenzoylhydrazone palladium(II) complexes as catalysts. *J. Chem. Soc. Dalt. Trans.* **1998**, 2715–2722. [CrossRef]
15. Kovala-Demertzi, D.; Yadav, P.N.; Demertzis, M.A.; Jasiski, J.P.; Andreadaki, F.J.; Kostas, I.D. First use of a palladium complex with a thiosemicarbazone ligand as catalyst precursor for the Heck reaction. *Tetrahedron Lett.* **2004**, *45*, 2923–2926. [CrossRef]
16. Kostas, I.D.; Andreadaki, F.J.; Kovala-Demertzi, D.; Christos, P.; Demertzis, M.A. Suzuki–Miyaura cross-coupling reaction of aryl bromides and chlorides with phenylboronic acid under aerobic conditions catalyzed by palladium complexes with thiosemicarbazone ligands. *Tetrahedron Lett.* **2005**, *46*, 1967–1970. [CrossRef]
17. Prajapati, N.P.; Patel, H.D. Novel thiosemicarbazone derivatives and their metal complexes: Recent development. *Synth. Commun.* **2019**, *49*, 1–38. [CrossRef]
18. Shakya, B.; Yadav, P.N. Thiosemicarbazones as potent anticancer agents and their modes of action. *Mini Rev. Med. Chem.* **2020**, *20*, 638–661. [CrossRef]
19. de Oliveira Carneiro Brum, J.; França, T.C.C.; LaPlante, S.R.; Villar, J.D.F. Synthesis and biological activity of hydrazones and derivatives: A review. *Mini Rev. Med. Chem.* **2020**, *20*, 342–368. [CrossRef]
20. Bonaccorso, C.; Marzo, T.; La Mendola, D. Biological applications of thiocarbohydrazones and their Metal complexes: A perspective review. *Pharmaceuticals* **2019**, *13*, 4. [CrossRef]
21. Namiecińska, E.; Sobiesiak, M.; Małecka, M.; Guga, P.; Rozalska, B.; Budzisz, E. Antimicrobial and structural properties of metal ions complexes with thiosemicarbazide motif and related heterocyclic compounds. *Curr. Med. Chem.* **2019**, *26*, 664–693. [CrossRef]
22. Summers, K.L. A structural chemistry perspective on the antimalarial properties of thiosemicarbazone metal complexes. *Mini Rev. Med. Chem.* **2019**, *19*, 569–590. [CrossRef]
23. Beraldo, H.; Gambino, D. The wide pharmacological versatility of semicarbazones, thiosemicarbazones and their metal complexes. *Mini Rev. Med. Chem.* **2004**, *4*, 31–39. [CrossRef]

24. Heffeter, P.; Pape, V.F.S.; Enyedy, É.A.; Keppler, B.K.; Szakacs, G.; Kowol, C.R. Anticancer thiosemicarbazones: Chemical properties, interaction with iron metabolism, and resistance development. *Antioxid. Redox Signal.* **2019**, *30*, 1062–1082. [CrossRef]
25. Mckenzie-Nickson, S.; Bush, A.I.; Barnham, K.J. Bis(thiosemicarbazone) metal complexes as therapeutics for neurodegenerative diseases. *Curr. Top. Med. Chem.* **2016**, *16*, 3058–3068. [CrossRef]
26. Matesanz, A.I.; Caballero, A.B.; Lorenzo, C.; Espargaró, A.; Sabaté, R.; Quiroga, A.G.; Gamez, P. Thiosemicarbazone derivatives as inhibitors of amyloid-β aggregation: Effect of metal coordination. *Inorg. Chem.* **2020**, *59*, 6978–6987. [CrossRef]
27. Brown, O.C.; Baguña Torres, J.; Holt, K.B.; Blower, P.J.; Went, M.J. Copper complexes with dissymmetrically substituted bis(thiosemicarbazone) ligands as a basis for PET radiopharmaceuticals: Control of redox potential and lipophilicity. *Dalt. Trans.* **2017**, *46*, 14612–14630. [CrossRef]
28. Dilworth, J.R.; Hueting, R. Metal complexes of thiosemicarbazones for imaging and therapy. *Inorganica Chim. Acta* **2012**, *389*, 3–15. [CrossRef]
29. Sharma, R.K.; Pandey, A.; Gulati, S.; Adholeya, A. An optimized procedure for preconcentration, determination and on-line recovery of palladium using highly selective diphenyldiketone-monothiosemicarbazone modified silica gel. *J. Hazard. Mater.* **2012**, *209–210*, 285–292. [CrossRef]
30. Bhaskar, R.; Sarveswari, S. Thiocarbohydrazide based Schiff base as a selective colorimetric and fluorescent chemosensor for Hg^{2+} with "turn-off" fluorescence responses. *Chem. Sel.* **2020**, *5*, 4050–4057. [CrossRef]
31. Udhayakumari, D.; Suganya, S.; Velmathi, S. Thiosemicabazone based fluorescent chemosensor for transition metal ions in aqueous medium. *J. Lumin.* **2013**, *141*, 48–52. [CrossRef]
32. Venkatachalam, T.K.; Bernhardt, P.V.; Pierens, G.K.; Stimson, D.H.R.; Bhalla, R.; Reutens, D.C. Synthesis and characterisation of indium(III) bis-thiosemicarbazone complexes: ^{18}F incorporation for PET imaging. *Aust. J. Chem.* **2019**, *72*, 383. [CrossRef]
33. Cortezon-Tamarit, F.; Sarpaki, S.; Calatayud, D.G.; Mirabello, V.; Pascu, S.I. Applications of "hot" and "cold" bis(thiosemicarbazonato) metal complexes in multimodal imaging. *Chem. Rec.* **2016**, *16*, 1380–1397. [CrossRef]
34. Singh, R.B.; Garg, B.S.; Singh, R.P. Analytical applications of thiosemicarbazones and semicarbazones: A review. *Talanta* **1978**, *25*, 619–632. [CrossRef]
35. Arion, V.B.; Platzer, S.; Rapta, P.; Machata, P.; Breza, M.; Vegh, D.; Dunsch, L.; Telser, J.; Shova, S.; Mac Leod, T.C.O.; et al. Marked stabilization of redox states and enhanced catalytic activity in galactose oxidase models based on transition metal S-methylisothiosemicarbazonates with −SR group in ortho position to the phenolic oxygen. *Inorg. Chem.* **2013**, *52*, 7524–7540. [CrossRef]
36. El-Hendawy, A.M.; Fayed, A.M.; Mostafa, M.R. Complexes of a diacetylmonoxime Schiff base of S-methyldithiocarbazate (H_2damsm) with Fe(III), Ru(III)/Ru(II), and V(IV); catalytic activity and X-ray crystal structure of [Fe(Hdamsm)$_2$]$NO_3 \cdot H_2O$. *Transit. Met. Chem.* **2011**, *36*, 351–361. [CrossRef]
37. Fayed, A.M.; Elsayed, S.A.; El-Hendawy, A.M.; Mostafa, M.R. Complexes of cis-dioxomolybdenum(VI) and oxovanadium(IV) with a tridentate ONS donor ligand: Synthesis, spectroscopic properties, X-ray crystal structure and catalytic activity. *Spectrochim. Acta Part A Mol. Biomol. Spectrosc.* **2014**, *129*, 293–302. [CrossRef]
38. Bjelogrlić, S.; Todorović, T.; Bacchi, A.; Zec, M.; Sladić, D.; Srdić-Rajić, T.; Radanović, D.; Radulović, S.; Pelizzi, G.; Anđelković, K. Synthesis, structure and characterization of novel Cd(II) and Zn(II) complexes with the condensation product of 2-formylpyridine and selenosemicarbazide. *J. Inorg. Biochem.* **2010**, *104*, 673–682. [CrossRef]
39. Gligorijević, N.; Todorović, T.; Radulović, S.; Sladić, D.; Filipović, N.; Gođevac, D.; Jeremić, D.; Anđelković, K. Synthesis and characterization of new Pt(II) and Pd(II) complexes with 2-quinolinecarboxaldehyde selenosemicarbazone: Cytotoxic activity evaluation of Cd(II), Zn(II), Ni(II), Pt(II) and Pd(II) complexes with heteroaromatic selenosemicarbazones. *Eur. J. Med. Chem.* **2009**, *44*, 1623–1629. [CrossRef]
40. Kowol, C.R.; Eichinger, R.; Jakupec, M.A.; Galanski, M.; Arion, V.B.; Keppler, B.K. Effect of metal ion complexation and chalcogen donor identity on the antiproliferative activity of 2-acetylpyridine N,N-dimethyl(chalcogen)semicarbazones. *J. Inorg. Biochem.* **2007**, *101*, 1946–1957. [CrossRef]
41. Kowol, C.R.; Reisner, E.; Chiorescu, I.; Arion, V.B.; Galanski, M.; Deubel, D.V.; Keppler, B.K. An electrochemical study of antineoplastic gallium, iron and ruthenium complexes with redox noninnocent α-N-heterocyclic chalcogensemicarbazones. *Inorg. Chem.* **2008**, *47*, 11032–11047. [CrossRef]

42. Molter, A.; Kaluđerović, G.N.; Kommera, H.; Paschke, R.; Langer, T.; Pöttgen, R.; Mohr, F. Synthesis, structures, [119]Sn Mössbauer spectroscopic studies and biological activity of some tin(IV) complexes containing pyridyl functionalised selenosemicarbazonato ligands. *J. Organomet. Chem.* **2012**, *701*, 80–86. [CrossRef]
43. Sarhan, A.M.; Elsayed, S.A.; Mashaly, M.M.; El-Hendawy, A.M. Oxovanadium(IV) and ruthenium(II) carbonyl complexes of ONS-donor ligands derived from dehydroacetic acid and dithiocarbazate: Synthesis, characterization, antioxidant activity, DNA binding and in vitro cytotoxicity. *Appl. Organomet. Chem.* **2019**, *33*, e4655. [CrossRef]
44. Shen, H.; Zhu, H.; Song, M.; Tian, Y.; Huang, Y.; Zheng, H.; Cao, R.; Lin, J.; Bi, Z.; Zhong, W. A selenosemicarbazone complex with copper efficiently down-regulates the 90-kDa heat shock protein HSP90AA1 and its client proteins in cancer cells. *BMC Cancer* **2014**, *14*, 629. [CrossRef]
45. Srdić-Rajić, T.; Zec, M.; Todorović, T.; Anđelković, K.; Radulović, S. Non-substituted N-heteroaromatic selenosemicarbazone metal complexes induce apoptosis in cancer cells via activation of mitochondrial pathway. *Eur. J. Med. Chem.* **2011**, *46*, 3734–3747. [CrossRef]
46. Todorović, T.R.; Vukašinović, J.; Portalone, G.; Suleiman, S.; Gligorijević, N.; Bjelogrlić, S.; Jovanović, K.; Radulović, S.; Anđelković, K.; Cassar, A.; et al. (Chalcogen)semicarbazones and their cobalt complexes differentiate HL-60 myeloid leukaemia cells and are cytotoxic towards tumor cell lines. *Medchemcomm* **2017**, *8*, 103–111. [CrossRef]
47. Zec, M.; Srdic-Rajic, T.; Krivokuca, A.; Jankovic, R.; Todorovic, T.; Andelkovic, K.; Radulovic, S. Novel selenosemicarbazone metal complexes exert anti-tumor effect via alternative, caspase-independent necroptotic cell death. *Med. Chem.* **2014**, *10*, 759–771. [CrossRef]
48. Mawat, T.H.; Al-Jeboori, M.J. Synthesis, characterisation, thermal properties and biological activity of coordination compounds of novel selenosemicarbazone ligands. *J. Mol. Struct.* **2020**, *1208*, 127876. [CrossRef]
49. Castle, T.C.; Maurer, R.I.; Sowrey, F.E.; Went, M.J.; Reynolds, C.A.; McInnes, E.J.L.; Blower, P.J. Hypoxia-targeting copper bis(selenosemicarbazone) complexes: Comparison with their sulfur analogues. *J. Am. Chem. Soc.* **2003**, *125*, 10040–10049. [CrossRef]
50. McQuade, P.; Martin, K.E.; Castle, T.C.; Went, M.J.; Blower, P.J.; Welch, M.J.; Lewis, J.S. Investigation into ^{64}Cu-labeled Bis(selenosemicarbazone) and Bis(thiosemicarbazone) complexes as hypoxia imaging agents. *Nucl. Med. Biol.* **2005**, *32*, 147–156. [CrossRef]
51. Dekanski, D.; Todorovic, T.; Mitic, D.; Filipovic, N.; Polovic, N.; Andjelkovic, K. High antioxidative potential and low toxic effects of selenosemicarbazone metal complexes. *J. Serbian Chem. Soc.* **2013**, *78*, 1503–1512. [CrossRef]
52. Djordjevic, I.; Vukasinovic, J.; Todorovic, T.; Filipovic, N.; Rodic, M.; Lolic, A.; Portalone, G.; Zlatovic, M.; Grubisic, S. Synthesis, structures and electronic properties of Co(III) complexes with 2-quinolinecarboxaldehyde thio- and selenosemicarbazone: A combined experimental and theoretical study. *J. Serb. Chem. Soc.* **2017**, *82*, 825–839. [CrossRef]
53. Campbell, M.J.M. Transition metal complexes of thiosemicarbazide and thiosemicarbazones. *Coord. Chem. Rev.* **1975**, *15*, 279–319. [CrossRef]
54. Casas, J.S.; García-Tasende, M.S.; Sordo, J. Main group metal complexes of semicarbazones and thiosemicarbazones. A structural review. *Coord. Chem. Rev.* **2000**, *209*, 197–261. [CrossRef]
55. Padhyé, S.; Kauffman, G.B. Transition metal complexes of semicarbazones and thiosemicarbazones. *Coord. Chem. Rev.* **1985**, *63*, 127–160. [CrossRef]
56. Selander, N.; Szabó, K.J. Catalysis by palladium pincer complexes. *Chem. Rev.* **2011**, *111*, 2048–2076. [CrossRef]
57. Phan, N.T.S.; Van Der Sluys, M.; Jones, C.W. On the nature of the active species in palladium catalyzed Mizoroki–Heck and Suzuki–Miyaura couplings–homogenous or heterogeneous catalysis, a critical review. *Adv. Synth. Catal.* **2006**, *348*, 609–679. [CrossRef]
58. Anitha, P.; Manikandan, R.; Endo, A.; Hashimoto, T.; Viswanathamurthi, P. Ruthenium(II) complexes containing quinone based ligands: Synthesis, characterization, catalytic applications and DNA interaction. *Spectrochim. Acta Part A Mol. Biomol. Spectrosc.* **2012**, *99*, 174–180. [CrossRef]
59. Asha, T.M.; Kurup, M.R.P. Synthesis, structural insights and catalytic activity of a dioxidomolybdenum(VI) complex chelated with N^4-(3-methoxyphenyl) thiosemicarbazone. *Transit. Met. Chem.* **2020**. [CrossRef]

60. Hosseini Monfared, H.; Kheirabadi, S.; Asghari Lalami, N.; Mayer, P. Dioxo- and oxovanadium(V) complexes of biomimetic hydrazone ONO and NNS donor ligands: Synthesis, crystal structure and catalytic reactivity. *Polyhedron* **2011**, *30*, 1375–1384. [CrossRef]
61. Islam, M.; Hossain, D.; Mondal, P.; Tuhina, K.; Roy, A.S.; Mondal, S.; Mobarak, M. Synthesis, characterization, and catalytic activity of a polymer-supported copper(II) complex with a thiosemicarbazone ligand. *Transit. Met. Chem.* **2011**, *36*, 223–230. [CrossRef]
62. Raja, G.; Sathya, N.; Jayabalakrishnan, C. Spectroscopic, catalytic, and biological studies on mononuclear ruthenium(II) ONSN chelating thiosemicarbazone complexes. *J. Coord. Chem.* **2011**, *64*, 817–831. [CrossRef]
63. Roy, S.; Saswati; Lima, S.; Dhaka, S.; Maurya, M.R.; Acharyya, R.; Eagle, C.; Dinda, R. Synthesis, structural studies and catalytic activity of a series of dioxidomolybdenum(VI)-thiosemicarbazone complexes. *Inorg. Chim. Acta* **2018**, *474*, 134–143. [CrossRef]
64. Thilagavathi, N.; Manimaran, A.; Padma Priya, N.; Sathya, N.; Jayabalakrishnan, C. Synthesis, spectroscopic, redox, catalytic and antimicrobial properties of some ruthenium(II) Schiff base complexes. *Transit. Met. Chem.* **2009**, *34*, 725–732. [CrossRef]
65. Ulaganatha Raja, M.; Gowri, N.; Ramesh, R. Synthesis, crystal structure and catalytic activity of ruthenium(II) carbonyl complexes containing ONO and ONS donor ligands. *Polyhedron* **2010**, *29*, 1175–1181. [CrossRef]
66. Wang, F.-M. Mononuclear oxovanadium(Iv) complex containing VO(ONS) basic core: Synthesis, structure, thermal gravimetric analysis, and catalytic property. *Inorg. Nano Met. Chem.* **2017**, *47*, 1380–1384. [CrossRef]
67. Maurya, M.R.; Dhaka, S.; Avecilla, F. Synthesis, characterization, reactivity and catalytic activity of dioxidomolybdenum(VI) complexes derived from tribasic ONS donor ligands. *Polyhedron* **2014**, *81*, 154–167. [CrossRef]
68. Maurya, A.; Kesharwani, N.; Kachhap, P.; Mishra, V.K.; Chaudhary, N.; Haldar, C. Polymer-anchored mononuclear and binuclear Cu^{II} Schiff-base complexes: Impact of heterogenization on liquid phase catalytic oxidation of a series of alkenes. *Appl. Organomet. Chem.* **2019**, *33*. [CrossRef]
69. Moradi-Shoeili, Z.; Boghaei, D.M.; Amini, M.; Bagherzadeh, M.; Notash, B. New molybdenum(VI) complex with ONS-donor thiosemicarbazone ligand: Preparation, structural characterization, and catalytic applications in olefin epoxidation. *Inorg. Chem. Commun.* **2013**, *27*, 26–30. [CrossRef]
70. Moradi-Shoeili, Z.; Zare, M. The effect of substituents on catalytic performance of bis-thiosemicarbazone Mo(VI) complexes: Synthesis and spectroscopic, electrochemical, and functional properties. *Kinet. Catal.* **2018**, *59*, 203–210. [CrossRef]
71. Muthukumar, M.; Viswanathamurthi, P. Synthesis, spectral characterization and catalytic studies of new ruthenium(II) chalcone thiosemicarbazone complexes. *Open Chem.* **2010**, *8*, 229–240. [CrossRef]
72. Nehar, O.K.; Mahboub, R.; Louhibi, S.; Roisnel, T.; Aissaoui, M. New thiosemicarbazone Schiff base ligands: Synthesis, characterization, catecholase study and hemolytic activity. *J. Mol. Struct.* **2020**, *1204*, 127566. [CrossRef]
73. Ngan, N.K.; Lo, K.M.; Wong, C.S.R. Dinuclear and polynuclear dioxomolybdenum(VI) Schiff base complexes: Synthesis, structural elucidation, spectroscopic characterization, electrochemistry and catalytic property. *Polyhedron* **2012**, *33*, 235–251. [CrossRef]
74. Nirmala, M.; Manikandan, R.; Prakash, G.; Viswanathamurthi, P. Ruthenium(II) complexes of hybrid 8-hydroxyquinoline-thiosemicarbazone ligands: Synthesis, characterization and catalytic applications. *Appl. Organomet. Chem.* **2014**, *28*, 18–26. [CrossRef]
75. Biswas, S.; Sarkar, D.; Roy, P.; Mondal, T.K. Synthesis and characterization of a ruthenium complex with bis(diphenylphosphino)propane and thioether containing ONS donor ligand: Application in transfer hydrogenation of ketones. *Polyhedron* **2017**, *131*, 1–7. [CrossRef]
76. Manikandan, R.; Anitha, P.; Prakash, G.; Vijayan, P.; Viswanathamurthi, P. Synthesis, spectral characterization and crystal structure of Ni(II) pyridoxal thiosemicarbazone complexes and their recyclable catalytic application in the nitroaldol (Henry) reaction in ionic liquid media. *Polyhedron* **2014**, *81*, 619–627. [CrossRef]
77. Raja, N.; Ramesh, R. Mononuclear ruthenium(III) complexes containing chelating thiosemicarbazones: Synthesis, characterization and catalytic property. *Spectrochim. Acta Part A Mol. Biomol. Spectrosc.* **2010**, *75*, 713–718. [CrossRef]
78. Barber, D.E.; Lu, Z.; Richardson, T.; Crabtree, R.H. Silane alcoholysis by a nickel(II) complex in a N, O, S ligand environment. *Inorg. Chem.* **1992**, *31*, 4709–4711. [CrossRef]

79. Anitha, P.; Manikandan, R.; Vijayan, P.; Anbuselvi, S.; Viswanathamurthi, P. Rhodium(I) complexes containing 9,10-phenanthrenequinone-*N*-substituted thiosemicarbazone ligands: Synthesis, structure, DFT study and catalytic diastereoselective nitroaldol reaction studies. *J. Organomet. Chem.* **2015**, *791*, 244–251. [CrossRef]
80. Anitha, P.; Viswanathamurthi, P.; Kesavan, D.; Butcher, R.J. Ruthenium(II) 9,10-phenanthrenequinone thiosemicarbazone complexes: Synthesis, characterization, and catalytic activity towards the reduction as well as condensation of nitriles. *J. Coord. Chem.* **2015**, *68*, 321–334. [CrossRef]
81. Manikandan, R.; Viswnathamurthi, P. Coordination behavior of ligand based on NNS and NNO donors with ruthenium(III) complexes and their catalytic and DNA interaction studies. *Spectrochim. Acta Part A Mol. Biomol. Spectrosc.* **2012**, *97*, 864–870. [CrossRef] [PubMed]
82. Youssef, N.S.; El-Zahany, E.; El-Seidy, A.M.A.; Caselli, A.; Fantauzzi, S.; Cenini, S. Synthesis and characterisation of new Schiff base metal complexes and their use as catalysts for olefin cyclopropanation. *Inorg. Chim. Acta* **2009**, *362*, 2006–2014. [CrossRef]
83. Youssef, N.S.; El-Seidy, A.M.A.; Schiavoni, M.; Castano, B.; Ragaini, F.; Gallo, E.; Caselli, A. Thiosemicarbazone copper complexes as competent catalysts for olefin cyclopropanations. *J. Organomet. Chem.* **2012**, *714*, 94–103. [CrossRef]
84. Kumar, A.; Kumar Rao, G.; Singh, A.K. Organochalcogen ligands and their palladium(II) complexes: Synthesis to catalytic activity for Heck coupling. *RSC Adv.* **2012**, *2*, 12552. [CrossRef]
85. Kumar, A.; Rao, G.K.; Kumar, S.; Singh, A.K. Organosulphur and related ligands in Suzuki–Miyaura C–C coupling. *Dalt. Trans.* **2013**, *42*, 5200. [CrossRef]
86. Kumar, A.; Rao, G.K.; Kumar, S.; Singh, A.K. Formation and role of palladium chalcogenide and other species in Suzuki–Miyaura and Heck C–C coupling reactions catalyzed with palladium(II) complexes of organochalcogen ligands: Realities and speculations. *Organometallics* **2014**, *33*, 2921–2943. [CrossRef]
87. Eremin, D.B.; Ananikov, V.P. Understanding active species in catalytic transformations: From molecular catalysis to nanoparticles, leaching, "cocktails" of catalysts and dynamic systems. *Coord. Chem. Rev.* **2017**, *346*, 2–19. [CrossRef]
88. Polynski, M.V.; Ananikov, V.P. Modeling Key pathways proposed for the formation and evolution of "cocktail"-type systems in Pd-catalyzed reactions involving ArX reagents. *ACS Catal.* **2019**, *9*, 3991–4005. [CrossRef]
89. Ortuño, M.A.; López, N. Reaction mechanisms at the homogeneous–heterogeneous frontier: Insights from first-principles studies on ligand-decorated metal nanoparticles. *Catal. Sci. Technol.* **2019**, *9*, 5173–5185. [CrossRef]
90. Trzeciak, A.M.; Augustyniak, A.W. The role of palladium nanoparticles in catalytic C–C cross-coupling reactions. *Coord. Chem. Rev.* **2019**, *384*, 1–20. [CrossRef]
91. Bourouina, A.; Meille, V.; de Bellefon, C. About solid phase vs. liquid phase in Suzuki-Miyaura reaction. *Catalysts* **2019**, *9*, 60. [CrossRef]
92. Sun, B.; Ning, L.; Zeng, H.C. Confirmation of Suzuki–Miyaura cross-coupling reaction mechanism through synthetic architecture of nanocatalysts. *J. Am. Chem. Soc.* **2020**, *142*, 13823–13832. [CrossRef] [PubMed]
93. Jagtap, S. Heck reaction—State of the art. *Catalysts* **2017**, *7*, 267. [CrossRef]
94. Xie, G.; Chellan, P.; Mao, J.; Chibale, K.; Smith, G.S. Thiosemicarbazone salicylaldiminato-palladium(II)-catalyzed Mizoroki-Heck reactions. *Adv. Synth. Catal.* **2010**, *352*, 1641–1647. [CrossRef]
95. Paul, P.; Datta, S.; Halder, S.; Acharyya, R.; Basuli, F.; Butcher, R.J.; Peng, S.-M.; Lee, G.-H.; Castiñeiras, A.; Drew, M.G.B.; et al. Syntheses, structures and efficient catalysis for C–C coupling of some benzaldehyde thiosemicarbazone complexes of palladium. *J. Mol. Catal. A Chem.* **2011**, *344*, 62–73. [CrossRef]
96. Prabhu, R.N.; Ramesh, R. Catalytic application of dinuclear palladium(II) bis(thiosemicarbazone) complex in the Mizoroki-Heck reaction. *Tetrahedron Lett.* **2012**, *53*, 5961–5965. [CrossRef]
97. Datta, S.; Seth, D.K.; Gangopadhyay, S.; Karmakar, P.; Bhattacharya, S. Nickel complexes of some thiosemicarbazones: Synthesis, structure, catalytic properties and cytotoxicity studies. *Inorganica Chim. Acta* **2012**, *392*, 118–130. [CrossRef]
98. Suganthy, P.K.; Prabhu, R.N.; Sridevi, V.S. Nickel(II) thiosemicarbazone complex catalyzed Mizoroki–Heck reaction. *Tetrahedron Lett.* **2013**, *54*, 5695–5698. [CrossRef]

99. Lima, J.C.; Nascimento, R.D.; Vilarinho, L.M.; Borges, A.P.; Silva, L.H.F.; Souza, J.R.; Dinelli, L.R.; Deflon, V.M.; da Hora Machado, A.E.; Bogado, A.L.; et al. Group 10 metal complexes with a tetradentate thiosemicarbazonate ligand: Synthesis, crystal structures and computational insights into the catalysis for C–C coupling via Mizoroki-Heck reaction. *J. Mol. Struct.* **2020**, *1199*, 126997. [CrossRef]
100. Kostas, I.D.; Heropoulos, G.A.; Kovala-Demertzi, D.; Yadav, P.N.; Jasinski, J.P.; Demertzis, M.A.; Andreadaki, F.J.; Vo-Thanh, G.; Petit, A.; Loupy, A. Microwave-promoted Suzuki–Miyaura cross-coupling of aryl halides with phenylboronic acid under aerobic conditions catalyzed by a new palladium complex with a thiosemicarbazone ligand. *Tetrahedron Lett.* **2006**, *47*, 4403–4407. [CrossRef]
101. Perreux, L.; Loupy, A. A tentative rationalization of microwave effects in organic synthesis according to the reaction medium, and mechanistic considerations. *Tetrahedron* **2001**, *57*, 9199–9223. [CrossRef]
102. Paul, P.; Bhattacharya, S. Organometallic complexes of the platinum metals: Synthesis, structure, and catalytic applications. *J. Chem. Sci.* **2012**, *124*, 1165–1173. [CrossRef]
103. Castiñeiras, A.; Fernandez-Hermida, N.; Garcia-Santos, I.; Gomez-Rodriguez, L. Neutral Ni^{II}, Pd^{II} and Pt^{II} ONS-pincer complexes of 5-acetylbarbituric-4N-dimethylthiosemicarbazone: Synthesis, characterization and properties. *Dalt. Trans.* **2012**, *41*, 13486–13495. [CrossRef] [PubMed]
104. Herrmann, W.A.; Reisinger, C.-P.; Öfele, K.; Broβmer, C.; Beller, M.; Fischer, H. Facile catalytic coupling of aryl bromides with terminal alkynes by phospha-palladacycles. *J. Mol. Catal. A Chem.* **1996**, *108*, 51–56. [CrossRef]
105. Ohff, M.; Ohff, A.; van der Boom, M.E.; Milstein, D. Highly active Pd(II) PCP-type catalysts for the Heck reaction. *J. Am. Chem. Soc.* **1997**, *119*, 11687–11688. [CrossRef]
106. Dutta, J.; Datta, S.; Kumar Seth, D.; Bhattacharya, S. Mixed-ligand benzaldehyde thiosemicarbazone complexes of palladium containing N,O-donor ancillary ligands. Syntheses, structures, and catalytic application in C–C and C–N coupling reactions. *RSC Adv.* **2012**, *2*, 11751. [CrossRef]
107. Dutta, J.; Bhattacharya, S. Controlled interaction of benzaldehyde thiosemicarbazones with palladium: Formation of bis-complexes with cis-geometry and organopalladium complexes, and their catalytic application in C–C and C–N coupling. *RSC Adv.* **2013**, *3*, 10707. [CrossRef]
108. Paul, P.; Sengupta, P.; Bhattacharya, S. Palladium mediated C–H bond activation of thiosemicarbazones: Catalytic application of organopalladium complexes in C–C and C–N coupling reactions. *J. Organomet. Chem.* **2013**, *724*, 281–288. [CrossRef]
109. Yan, H.; Chellan, P.; Li, T.; Mao, J.; Chibale, K.; Smith, G.S. Cyclometallated Pd(II) thiosemicarbazone complexes: New catalyst precursors for Suzuki-coupling reactions. *Tetrahedron Lett.* **2013**, *54*, 154–157. [CrossRef]
110. Pandiarajan, D.; Ramesh, R.; Liu, Y.; Suresh, R. Palladium(II) thiosemicarbazone-catalyzed Suzuki–Miyaura cross-coupling reactions of aryl halides. *Inorg. Chem. Commun.* **2013**, *33*, 33–37. [CrossRef]
111. Verma, P.R.; Mandal, S.; Gupta, P.; Mukhopadhyay, B. Carbohydrate derived thiosemicarbazone and semicarbazone palladium complexes: Homogeneous catalyst for C–C cross coupling reactions. *Tetrahedron Lett.* **2013**, *54*, 4914–4917. [CrossRef]
112. Tenchiu, A.-C.; Ventouri, I.-K.; Ntasi, G.; Palles, D.; Kokotos, G.; Kovala-Demertzi, D.; Kostas, I.D. Synthesis of a palladium complex with a β-D-glucopyranosyl-thiosemicarbazone and its application in the Suzuki–Miyaura coupling of aryl bromides with phenylboronic acid. *Inorg. Chim. Acta* **2015**, *435*, 142–146. [CrossRef]
113. Matsinha, L.C.; Mao, J.; Mapolie, S.F.; Smith, G.S. Water-soluble palladium(II) sulfonated thiosemicarbazone complexes: Facile synthesis and preliminary catalytic studies in the Suzuki-Miyaura cross-coupling reaction in water. *Eur. J. Inorg. Chem.* **2015**, *2015*, 4088–4094. [CrossRef]
114. Baruah, J.; Gogoi, R.; Gogoi, N.; Borah, G. A thiosemicarbazone–palladium(II)–imidazole complex as an efficient pre-catalyst for Suzuki–Miyaura cross-coupling reactions at room temperature in aqueous media. *Transit. Met. Chem.* **2017**, *42*, 683–692. [CrossRef]
115. Dharani, S.; Kalaiarasi, G.; Sindhuja, D.; Lynch, V.M.; Shankar, R.; Karvembu, R.; Prabhakaran, R. Tetranuclear palladacycles of 3-acetyl-7-methoxy-2H-chromen-2-one derived Schiff bases: Efficient catalysts for Suzuki–Miyaura coupling in an aqueous medium. *Inorg. Chem.* **2019**, *58*, 8045–8055. [CrossRef] [PubMed]
116. Bakir, M.; Lawrence, M.W.; Bohari Yamin, M. Novel κ^2-N$_{im}$,S- and κ^4-C,N$_{im}$,(μ-S),(μ-S)-coordination of di-2-thienyl ketone thiosemicarbazone (dtktsc). Hydrogen evolution and catalytic properties of palladacyclic [Pd(κ^4-C,N$_{im}$,(μ-S),(μ-S)-dtktsc-2H)]$_4$. *Inorg. Chim. Acta* **2020**, *507*, 119592. [CrossRef]

117. Thapa, K.; Paul, P.; Bhattacharya, S. A group of diphosphine-thiosemicarbazone complexes of palladium: Efficient precursors for catalytic C–C and C–N coupling reactions. *Inorg. Chim. Acta* **2019**, *486*, 232–239. [CrossRef]
118. Datta, S.; Seth, D.K.; Butcher, R.J.; Bhattacharya, S. Mixed-ligand thiosemicarbazone complexes of nickel: Synthesis, structure and catalytic activity. *Inorg. Chim. Acta* **2011**, *377*, 120–128. [CrossRef]
119. Anitha, P.; Manikandan, R.; Vijayan, P.; Prakash, G.; Viswanathamurthi, P.; Butcher, R.J. Nickel(II) complexes containing ONS donor ligands: Synthesis, characterization, crystal structure and catalytic application towards C–C cross-coupling reactions. *J. Chem. Sci.* **2015**, *127*, 597–608. [CrossRef]
120. Prabhu, R.N.; Ramesh, R. Synthesis and structural characterization of Pd(II) thiosemicarbazonato complex: Catalytic evaluation in synthesis of diaryl ketones from aryl aldehydes and arylboronic acids. *Tetrahedron Lett.* **2017**, *58*, 405–409. [CrossRef]
121. Sonogashira, K.; Tohda, Y.; Hagihara, N. A convenient synthesis of acetylenes: Catalytic substitutions of acetylenic hydrogen with bromoalkenes, iodoarenes and bromopyridines. *Tetrahedron Lett.* **1975**, *16*, 4467–4470. [CrossRef]
122. Prabhu, R.N.; Pal, S. Copper-free Sonogashira reactions catalyzed by a palladium(II) complex bearing pyrenealdehyde thiosemicarbazonate under ambient conditions. *Tetrahedron Lett.* **2015**, *56*, 5252–5256. [CrossRef]
123. Prabhu, R.N.; Ramesh, R. Square-planar Ni(II) thiosemicarbazonato complex as an easily accessible and convenient catalyst for Sonogashira cross-coupling reaction. *Tetrahedron Lett.* **2016**, *57*, 4893–4897. [CrossRef]
124. Lu, L.; Chellan, P.; Smith, G.S.; Zhang, X.; Yan, H.; Mao, J. Thiosemicarbazone salicylaldiminato palladium(II)-catalyzed alkynylation couplings between arylboronic acids and alkynes or alkynyl carboxylic acids. *Tetrahedron* **2014**, *70*, 5980–5985. [CrossRef]
125. Rokade, B.V.; Barker, J.; Guiry, P.J. Development of and recent advances in asymmetric A3 coupling. *Chem. Soc. Rev.* **2019**, *48*, 4766–4790. [CrossRef]
126. Manikandan, R.; Anitha, P.; Viswanathamurthi, P.; Malecki, J.G. Palladium(II) pyridoxal thiosemicarbazone complexes as efficient and recyclable catalyst for the synthesis of propargylamines by a three component coupling reactions in ionic liquids. *Polyhedron* **2016**, *119*, 300–306. [CrossRef]
127. Heravi, M.M.; Hashemi, E.; Nazari, N. Negishi coupling: An easy progress for C–C bond construction in total synthesis. *Mol. Divers.* **2014**, *18*, 441–472. [CrossRef]
128. Muthukumar, M.; Sivakumar, S.; Viswanathamurthi, P.; Karvembu, R.; Prabhakaran, R.; Natarajan, K. Studies on ruthenium(III) chalcone thiosemicarbazone complexes as catalysts for carbon–carbon coupling. *J. Coord. Chem.* **2010**, *63*, 296–306. [CrossRef]
129. Priyarega, S.; Kalaivani, P.; Prabhakaran, R.; Hashimoto, T.; Endo, A.; Natarajan, K. Nickel(II) complexes containing thiosemicarbazone and triphenylphosphine: Synthesis, spectroscopy, crystallography and catalytic activity. *J. Mol. Struct.* **2011**, *1002*, 58–62. [CrossRef]
130. Güveli, Ş.; Agopcan Çınar, S.; Karahan, Ö.; Aviyente, V.; Ülküseven, B. Nickel(II)-PPh$_3$ complexes of S, N-substituted thiosemicarbazones-structure, DFT study, and catalytic efficiency. *Eur. J. Inorg. Chem.* **2016**, *2016*, 538–544. [CrossRef]
131. Dorel, R.; Grugel, C.P.; Haydl, A.M. The Buchwald–Hartwig amination after 25 years. *Angew. Chem. Int. Ed.* **2019**, *58*, 17118–17129. [CrossRef]
132. Heravi, M.M.; Kheilkordi, Z.; Zadsirjan, V.; Heydari, M.; Malmir, M. Buchwald-Hartwig reaction: An overview. *J. Organomet. Chem.* **2018**, *861*, 17–104. [CrossRef]
133. Munir, I.; Zahoor, A.F.; Rasool, N.; Naqvi, S.A.R.; Zia, K.M.; Ahmad, R. Synthetic applications and methodology development of Chan–Lam coupling: A review. *Mol. Divers.* **2019**, *23*, 215–259. [CrossRef] [PubMed]
134. West, M.J.; Fyfe, J.W.B.; Vantourout, J.C.; Watson, A.J.B. Mechanistic development and recent applications of the Chan–Lam amination. *Chem. Rev.* **2019**, *119*, 12491–12523. [CrossRef]
135. Khan, F.; Dlugosch, M.; Liu, X.; Banwell, M.G. The palladium-catalyzed Ullmann cross-coupling reaction: A modern variant on a time-honored process. *Acc. Chem. Res.* **2018**, *51*, 1784–1795. [CrossRef]
136. Jiang, J.; Du, L.; Ding, Y. Aryl-aryl bond formation by Ullmann Reaction: From mechanistic aspects to catalyst. *Mini. Rev. Org. Chem.* **2020**, *17*, 26–46. [CrossRef]

137. Prabhu, R.N.; Ramesh, R. Synthesis and structural characterization of palladium(II) thiosemicarbazone complex: Application to the Buchwald–Hartwig amination reaction. *Tetrahedron Lett.* **2013**, *54*, 1120–1124. [CrossRef]
138. Anitha, P.; Manikandan, R.; Viswanathamurthi, P. Palladium(II) 9,10-phenanthrenequinone N-substituted thiosemicarbazone/semicarbazone complexes as efficient catalysts for N-arylation of imidazole. *J. Coord. Chem.* **2015**, *68*, 3537–3550. [CrossRef]
139. Shan, Y.; Wang, Y.; Jia, X.; Cai, W.; Xiang, J. New, simple, and effective thiosemicarbazide ligand for copper(II)-catalyzed N-arylation of imidazoles. *Synth. Commun.* **2012**, *42*, 1192–1199. [CrossRef]
140. Gogoi, N.; Borah, G.; Gogoi, P.K. Cu(II) complex of phenylthiosemicarbazone: An *in situ* catalyst for formation of C–N bond between different N-based neucleophiles with arylboronic acids at room temperature. *Heteroat. Chem.* **2018**, *29*, e21414. [CrossRef]
141. Ramachandran, R.; Prakash, G.; Vijayan, P.; Viswanathamurthi, P.; Grzegorz Malecki, J. Synthesis of heteroleptic copper(I) complexes with phosphine-functionalized thiosemicarbazones: An efficient catalyst for regioselective N-alkylation reactions. *Inorg. Chim. Acta* **2017**, *464*, 88–93. [CrossRef]
142. Ramachandran, R.; Prakash, G.; Selvamurugan, S.; Viswanathamurthi, P.; Malecki, J.G.; Ramkumar, V. Efficient and versatile catalysis of N-alkylation of heterocyclic amines with alcohols and one-pot synthesis of 2-aryl substituted benzazoles with newly designed ruthenium(II) complexes of PNS thiosemicarbazones. *Dalt. Trans.* **2014**, *43*, 7889–7902. [CrossRef]
143. Ramachandran, R.; Prakash, G.; Nirmala, M.; Viswanathamurthi, P.; Malecki, J.G. Ruthenium(II) carbonyl complexes designed with arsine and PNO/PNS ligands as catalysts for N-alkylation of amines via hydrogen autotransfer process. *J. Organomet. Chem.* **2015**, *791*, 130–140. [CrossRef]
144. Ramachandran, R.; Prakash, G.; Selvamurugan, S.; Viswanathamurthi, P.; Malecki, J.G.; Linert, W.; Gusev, A. Ruthenium(II) complexes containing a phosphine-functionalized thiosemicarbazone ligand: Synthesis, structures and catalytic C–N bond formation reactions via N-alkylation. *RSC Adv.* **2015**, *5*, 11405–11422. [CrossRef]
145. Ramachandran, R.; Prakash, G.; Viswanathamurthi, P.; Malecki, J.G. Ruthenium(II) complexes containing phosphino hydrazone/thiosemicarbazone ligand: An efficient catalyst for regioselective N-alkylation of amine via borrowing hydrogen methodology. *Inorg. Chim. Acta* **2018**, *477*, 122–129. [CrossRef]
146. Suganthy, P.K.; Prabhu, R.N.; Sridevi, V.S. Palladium(II) thiosemicarbazone complex: Synthesis, structure and application to carbon–oxygen cross-coupling reaction. *Inorg. Chem. Commun.* **2014**, *44*, 67–69. [CrossRef]
147. Hübner, S.; de Vries, J.G.; Farina, V. Why does industry not use immobilized transition metal complexes as catalysts? *Adv. Synth. Catal.* **2016**, *358*, 3–25. [CrossRef]
148. Kovala-Demertzi, D.; Kourkoumelis, N.; Derlat, K.; Michalak, J.; Andreadaki, F.J.; Kostas, I.D. Thiosemicarbazone-derivatised palladium nanoparticles as efficient catalyst for the Suzuki-Miyaura cross-coupling of aryl bromides with phenylboronic acid. *Inorg. Chim. Acta* **2008**, *361*, 1562–1565. [CrossRef]
149. Bakherad, M.; Keivanloo, A.; Bahramian, B.; Jajarmi, S. Synthesis of ynones via recyclable polystyrene-supported palladium(0) complex catalyzed acylation of terminal alkynes with acyl chlorides under copper- and solvent-free conditions. *Synlett* **2011**, *2011*, 311–314. [CrossRef]
150. Sharma, R.K.; Pandey, A.; Gulati, S. Silica-supported palladium complex: An efficient, highly selective and reusable organic–inorganic hybrid catalyst for the synthesis of E-stilbenes. *Appl. Catal. A Gen.* **2012**, *431–432*, 33–41. [CrossRef]
151. Veisi, H.; Metghalchi, Y.; Hekmati, M.; Samadzadeh, S. Cu^I heterogenized on thiosemicarbazide modified-multi walled carbon nanotubes (thiosemicarbazide-MWCNTs-Cu^I): Novel heterogeneous and reusable nanocatalyst in the C–N Ullmann coupling reactions. *Appl. Organomet. Chem.* **2017**, *31*, e3676. [CrossRef]
152. Sadjadi, S. Palladium nanoparticles immobilized on cyclodextrin-decorated halloysite nanotubes: Efficient heterogeneous catalyst for promoting copper- and ligand-free Sonogashira reaction in water–ethanol mixture. *Appl. Organomet. Chem.* **2018**, *32*, e4211. [CrossRef]
153. Sadjadi, S.; Hosseinnejad, T.; Malmir, M.; Heravi, M.M. Cu@furfural imine-decorated halloysite as an efficient heterogeneous catalyst for promoting ultrasonic-assisted A3 and KA2 coupling reactions: A combination of experimental and computational study. *New J. Chem.* **2017**, *41*, 13935–13951. [CrossRef]

154. Sadjadi, S.; Heravi, M.M.; Ebrahimizadeh, M. Synthesis of Cu@Fur-SBA-15 as a novel efficient and heterogeneous catalyst for promoting A3-coupling under green and mild reaction conditions. *J. Porous Mater.* **2018**, *25*, 779–788. [CrossRef]
155. Ghodsinia, S.S.E.; Akhlaghinia, B. Cu^I anchored onto mesoporous SBA-16 functionalized by aminated 3-glycidyloxypropyltrimethoxysilane with thiosemicarbazide (SBA-16/GPTMS-TSC-Cu^I): A heterogeneous mesostructured catalyst for S-arylation reaction under solvent-free conditions. *Green Chem.* **2019**, *21*, 3029–3049. [CrossRef]
156. Ahmadi, A.; Sedaghat, T.; Azadi, R.; Motamedi, H. Magnetic mesoporous silica nanocomposite functionalized with palladium Schiff base complex: Synthesis, characterization, catalytic efficacy in the Suzuki–Miyaura reaction and α-amylase immobilization. *Catal. Lett.* **2020**, *150*, 112–126. [CrossRef]

© 2020 by the authors. Licensee MDPI, Basel, Switzerland. This article is an open access article distributed under the terms and conditions of the Creative Commons Attribution (CC BY) license (http://creativecommons.org/licenses/by/4.0/).

Review

Recent Advances in Metal-Catalyzed Alkyl–Boron (C(sp^3))–C(sp^2)) Suzuki-Miyaura Cross-Couplings

Janwa El-Maiss [1,†], Tharwat Mohy El Dine [2,†], Chung-Shin Lu [3], Iyad Karamé [4], Ali Kanj [4], Kyriaki Polychronopoulou [5,6,*] and Janah Shaya [7,8,*]

[1] Université d'Orléans, Institut de Chimie Organique et Analytique, Pole Chimie, Rue de Chartres, 45000 Orléans, France; janwa.el-maiss@univ-orleans.fr
[2] Institut des Biomolécules Max Mousseron (IBMM), UMR 5247, École nationale supérieure de chimie de Montpellier (ENSCM), 240 Avenue du Professeur Emile Jeanbrau, 34090 Montpellier, France; Tharwat.Mohy-El-Dine@enscm.fr
[3] Department of General Education, National Taichung University of Science and Technology, Taichung 404, Taiwan; cslu6@nutc.edu.tw
[4] Laboratory of Catalysis Organometallic and Materials (LCOM), Depart. of Chemistry, Faculty of Science I, Lebanese University-Hariri Campus, Hadath, Lebanon; iyad.karameh@ul.edu.lb (I.K.); kani@ul.edu.lb (A.K.)
[5] Department of Mechanical Engineering, Khalifa University of Science and Technology, Abu Dhabi, P.O. Box 127788, UAE
[6] Center for Catalysis and Separation, Khalifa University of Science and Technology, Abu Dhabi, P.O. Box 127788, UAE
[7] College of Arts and Sciences, Khalifa University, Abu Dhabi, P.O. Box 127788, UAE
[8] College of Medicine and Health Sciences, Khalifa University, Abu Dhabi, P.O. Box 127788, UAE
* Correspondence: kyriaki.polychrono@ku.ac.ae (K.P.); shaya.janah@ku.ac.ae (J.S.)
† J.E.-M. and T.M.E.D. contributed equally to this work and shall be considered Co-First authors.

Received: 25 January 2020; Accepted: 19 February 2020; Published: 5 March 2020

Abstract: Boron chemistry has evolved to become one of the most diverse and applied fields in organic synthesis and catalysis. Various valuable reactions such as hydroborylations and Suzuki–Miyaura cross-couplings (SMCs) are now considered as indispensable methods in the synthetic toolbox of researchers in academia and industry. The development of novel sterically- and electronically-demanding C(sp^3)–Boron reagents and their subsequent metal-catalyzed cross-couplings attracts strong attention and serves in turn to expedite the wheel of innovative applications of otherwise challenging organic adducts in different fields. This review describes the significant progress in the utilization of classical and novel C(sp^3)–B reagents (9-BBN and 9-MeO-9-BBN, trifluoroboronates, alkylboranes, alkylboronic acids, MIDA, etc.) as coupling partners in challenging metal-catalyzed C(sp^3)–C(sp^2) cross-coupling reactions, such as B-alkyl SMCs after 2001.

Keywords: Suzuki–Miyaura cross-couplings; C(sp^3) –C(sp^2); alkylboron reagents; metal catalysis

---

1. Introduction

Boron is a peculiar metalloid with fascinating chemical complexity. The unusual properties of boron stem from its three valence electrons, which can be easily torn away, favoring metallicity and making it electron-deficient, yet sufficiently localized and tightly bound to the nucleus, consequently allowing the insulating states to emerge [1]. Boron compounds have been intensively investigated for energy storage applications, particularly due to the relatively low atomic mass of boron (10.811 ± 0.007 amu). The energy-related uses of boron compounds range from high-energy fuels for advanced aircrafts to boron–nitrogen–hydrogen compounds as hydrogen storage materials for fuel cells [2]. The rich pioneering research on boron resulted in the consecutive awarding of two Nobel Prizes in chemistry in 1976 and 1979 [3,4].

Organoboron compounds (e.g., boronic acids, boronic esters and boronamides) generally comprise at least one carbon–boron (C–B) bond (Scheme 1A) [5–8]. Organoboron compounds were initially used in organic synthesis 60 years ago [9,10]. Ever since, chemistries involving such compounds continued to advance until these reagents have become one of the most diverse, widely studied and applied families in catalysis and organic synthesis [10,11]. Currently, they are engaged in numerous classic and important reactions such as hydroborations and Suzuki–Miyaura cross-couplings (SMCs), among others [8]. The SMC reaction generally involves the conjoining of an organoboron reagent and an organic halide or pseudohalide in the presence of palladium (or other relevant metal/ligand) as a catalyst and a base for the activation of the boron compound (Scheme 1B) [5–7,12]. Organoboron compounds have also found several applications in pharmaceuticals where boron-based drugs exemplify a novel class of molecules for several biomedical applications as molecular imaging agents (optical/nuclear imaging) and neutron capture therapy agents (BNCT), as well as therapeutic agents (anticancer, antiviral, antibacterial, etc.) [13]. Likewise, the utility and ubiquity of boron-based compounds have bolstered the development of agricultural and material sciences [14,15]. Organoborane polymers have been investigated as electrolytes for batteries, electro-active materials, and supported Lewis acid catalysts [16,17].

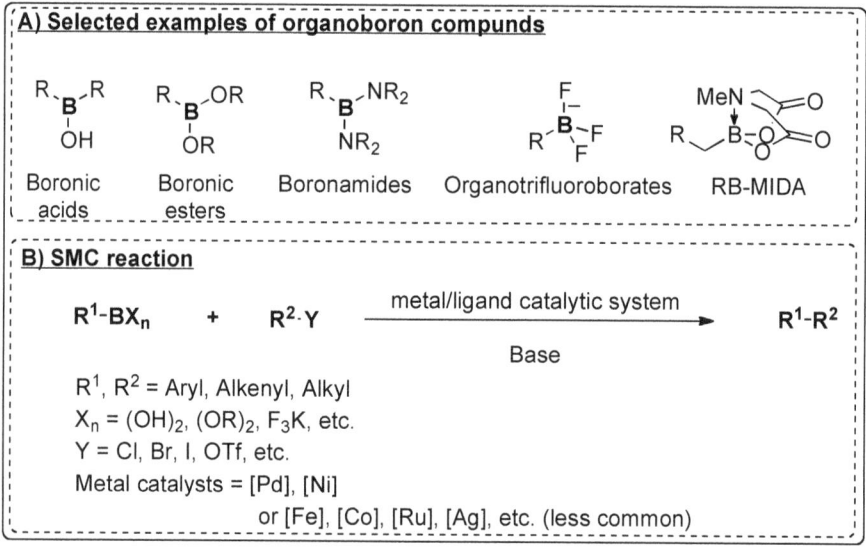

Scheme 1. (A) Examples of organoboron compounds, (B) Suzuki–Miyaura cross-coupling reaction.

Metal catalysis has had a major impact on numerous research fields from energy, biomass, environmental and water purification to synthesis of otherwise challenging and even inaccessible materials and medicinal adducts [18–30]. In line, the intensive research in metal catalysis has led to significant progress in borylation of primary $C(sp^3)$–H bonds of unfunctionalized hydrocarbons, allowing access to a variety of $C(sp^3)$–B reagents and consequent breakthroughs in $C(sp^3)$–$C(sp,sp^2,sp^3)$ cross-couplings. Comprehensive work has been done on the development of an efficient sp^2–sp^2 SMC; however, there have been far fewer reports on sp^3–sp^2 or sp^3–sp^3 variants [31–38]. Among the different hybridized boron reagents employed in SMCs (e.g., aryl, heteroaryl, and vinylboronic acids and esters), the use of organoboron compounds with alkyl groups (sp^3 carbon) was severely limited in these coupling reactions due to competitive side reactions [39,40]. Organometallic compounds that are metalated at sp^3 carbon atoms and especially containing β-hydrogen atoms give rise to alkyl–palladium complexes that are susceptible to β-hydride elimination rather than reductive elimination [41]. Furthermore, although boronic acids are relatively stable at ambient temperature and can be isolated by chromatography and crystallization, they favor other side reactions such as protodeboronation under SMC conditions [42]. The undesired decomposition pathways in sp^3–boron couplings are mostly circumvented by using tetrahedral boronates (e.g., potassium trifluoroborates (RBF$_3$K) and N-methyliminodiacetyl boronates (RB–[MIDA]; Scheme 1A) or stoichiometric loadings of palladium catalysts. On the other hand, the use of alkylborane (B-alkyl-9-borabicyclo [3.3.1]nonane: B-alkyl-9-BBN) in sp^3 SMCs suffers from isolation difficulties, lack of atom economy, air sensitivity and functional group tolerance (e.g., to ketones). Trialkylboranes (R$_3$B) have also been employed in SMCs [43,44].

The alkyl–alkyl SMCs (sp^3–sp^3) were recently reviewed in 2017 [45]. Hence, we will focus here on the recent development in cross-coupling reactions using sp^3–boron reagents and C(sp^2)–reagents. One class of the sp^3–sp^2 SMC is commonly known as *B–alkyl Suzuki–Miyaura cross-coupling*. It is distinguished from the other SMCs in that this cross-coupling occurs between an alkyl borane and an aryl or vinyl halide, triflate or enol phosphate. Generally, the most reactive partners for B–alkyl SMC are unhindered electron-rich organoboranes and electron-deficient coupling partners (halides or triflates). Notably, this type of coupling is highly affected by all the reaction parameters including the type of organoborane, base, solvent and metal catalyst, and the nature of the halide partner. The effects of these parameters were detailed in the review by Danishefsky et al. on B–alkyl SMC in 2001 [33]. This work will thus summarize the C(sp^3)–C(sp^2) cross-couplings covering the more recent progress in this area after 2001. The advances in stereospecific sp^3–sp^2 SMCs will be out of the scope of this highlight. However, it is worth noting that different versions that proceed with either retention or inversion of configuration have been well established [46,47]. Acyl SMC (acid halides, anhydrides, amides, esters), decarbonylative SMC and Liebeskind–Srogl cross-couplings are also not covered here and were recently reviewed in the literature extensively [48–52].

2. Suzuki–Miyaura Cross-Coupling (SMC)

As mentioned in the introduction, SMC is the conjoining of an organoboron reagent and an organic halide or pseudohalide in the presence of palladium (or other relevant metal) as a catalyst and a base for the activation of the boron compound (Scheme 1B) [5–7]. The efficiency of palladium has contributed to the ever-accelerating advances in catalysis, where coupling reactions, including SMC ones, are nowadays performed at ppb (parts per billion) molar catalyst loadings [53]. Nickel has also proved to have an efficient catalytic activity for SMC as the expensive palladium catalysts [54,55]. The high reactivity of nickel was revealed with difficult substrates such as aryl chlorides/mesylates, whose coupling reactions do not proceed easily with conventional Pd catalysis. In addition to being inexpensive, nickel catalysts can be more easily removed from the reaction mixtures while their economic practicality eliminates the need to recycle them [56]. Other metal catalytic systems have been investigated in SMC reactions such as Fe, Co, Ru, Cu, Ag, etc. However, their applications are by far less than Pd and Ni catalysts [56–58].

Since its discovery in 1979 [59], the Suzuki–Miyaura reaction has arguably become one of the most widely-applied, simple and versatile transition metal-catalyzed methods used for the construction of C–C bonds [60]. The general catalytic cycle is similar to other metal-catalyzed cross-couplings starting with an oxidative addition followed by a transmetalation and ending with a reductive elimination (Scheme 2). Transmetalation or the activation of the boron reagent makes Suzuki–Miyaura coupling different than other transition-metal cross-couplings processes. Mechanistic investigations were able to illustrate the role of each reagent in the reaction medium in addition to the metal. Some insights are now well established such as the necessity of sigma-rich electron-donor ligands, protic solvents and the base [61,62]; other mechanistic insights are still active areas of research including the activation way of boron in presence of the base. Two main analysis routes can be outlined as can be seen in Scheme 2: A) Boronate pathway: tetracoordinate nucleophilic boronate species **III** is generated *in situ* and substitutes the halide ligand of the Pd intermediate **I** issued from the oxidative addition, followed by the elimination of B(OH)$_2$OR from the resulting intermediate **IV** to transfer the organic moiety to palladium species **V**. B) Oxo-palladium pathway: the RO$^-$ substitute ligand X on the palladium center leading to oxo-palladium **II** which acts as a nucleophile toward the boronic acid species, generating the tetracoordinate species **IV**. Ambiguity occurs since inorganic bases in aqueous or alcohol solvents, generating the required alkoxy or hydroxy ligands, are commonly employed in the SMC, to accelerate either pathway **A** or **B**. However, all DFT (Density Functional Theory) studies and ES-MS (Electrospray Ionization-Mass Spectrometry) investigations [63,64], where boronate species were observed and not oxo-palladium ones [65–67], support pathway **A**. Studies defending the suggestion of pathway **B** consist of kinetic analysis and experimental observations of the lack of activities in some cases in the presence of organic Lewis bases or lithium salts of boronic species. The group of Maseras claimed that while pathway **A** and pathway **B** are competitive, the first has lower energy barriers than the second [68]. Therefore, the boronate pathway (**A**) is faster. Additionally, they stated that their theoretical report is consistent with the experimental observations they reproduced [63].

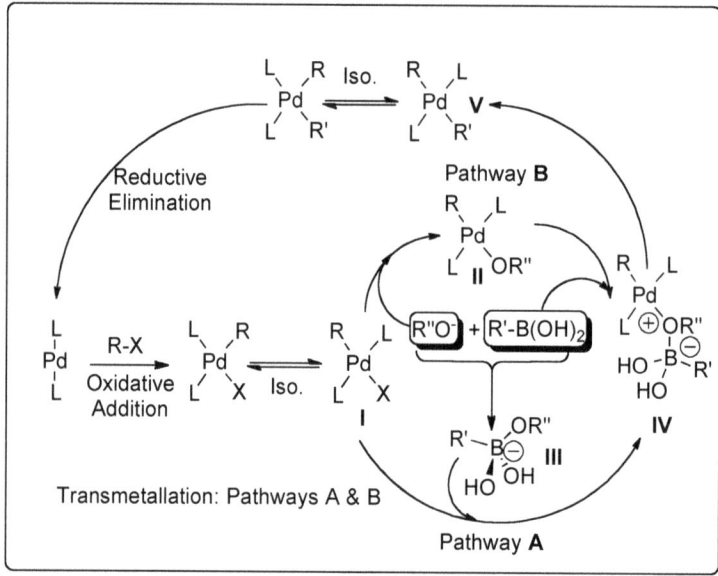

Scheme 2. General mechanism of Suzuki–Miyaura cross-coupling.

Further investigation is needed to conclude which pathway is the actual one, or whether both exist in a competitive manner in each catalytic cycle. One point supporting pathway **A** can still be considered here. The formation of oxo palladium **II** is less favored in the case where the palladium center is electron-rich (bearing a good sigma donor and weak π acceptor ligands), which is more likely to react with a weaker nucleophile like boronate [R–B(OH)$_3$]$^-$ rather than with a strong nucleophile, such as hydroxy or alkoxy groups.

The success of the SMC method originates from its high regio- and stereo-selectivity, extremely low catalytic loadings, and the exceptionally mild reaction conditions. The employed conditions are compatible with aqueous and heterogeneous media and tolerate steric hindrance and a wide range of functional groups. In addition, the readily available organoboron reagents and the versatile developed methods that permit access to challenging boron-functionalized adducts as well as the easy incorporation of nontransferable boron ligands have contributed to the appeal of SMC reactions. Most boron starting materials are thermally stable and inert to oxygen, water and related solvents. In general, they are relatively non-toxic and environmentally benign, and so are their by-products. Thus, they can be handled and separated easily from the reaction mixtures [69–72]. These unique features have allowed researchers to utilize SMC in a great variety of applications from development of polymeric materials to total synthesis of complex natural products. SMCs also constitute an important tool in medicinal chemistry, in production of fine chemicals and innovative materials as organic-light emitting diodes and in large-scale syntheses of pharmaceuticals [73–75]. Several reviews and textbooks have been dedicated to the applications of SMCs. Our review will only highlight a few examples of target molecules in Section 9. The relevant reports of $C(sp^3)$–$C(sp^2)$ cross-couplings are summarized in Table 1 (*reaction partners and conditions*) in order of the respective sections where they are discussed.

Table 1. General summary of the relevant reports of $C(sp^3)$–$C(sp^2)$ cross-couplings in this review.

Boron Reagent	Substrate	Reaction Conditions (General)	Reference	Section and Scheme
B-alkyl-9-BBN and trialkylboranes	Aryl iodides	$PdCl_2(dppf)$, NaOH, THF, reflux, 16 h	76	3;3A
Alkylboranes	Aryl bromides and iodides	$PdCl_2(dppf)$, NaOH, THF, 65 °C	77–79	3;3B
B-alkyl-9-BBN and boronic acids	Aryl halides	$Pd(OAc)_2$, SPhos, $K_3PO_4 \cdot H_2O$, THF or toluene	80	3;3C
B-benzyl-9-BBN	Chloroenynes	$Pd(PPh_3)_4$, Cs_2CO_3, water, 60 °C, 12 h	81	3;4
B-alkyl-9-BBN	C_{Ar}-O electrophiles	$Ni(COD)_2$, IPr.HCl, Cs_sCO_3, iPr_2O, 110 °C, 12 h	85	3;5A
B-alkyl-9-BBN	Aromatic and alkenyl ethers	$Ni(COD)_2$, PCy_3, base, iPr_2O, 110 °C	86	3;5B,C
1,3-dienes and 9-BBN	Aryl halides	$Pd(dppf)Cl_2$ or $Pd(dppb)Cl_2$, NaOH, THF, 40 or 65 °C	87	3;6
B-alkyl-9-BBN	β-triflyl enones	$Pd(dppf)Cl_2$, Cs_2CO_3, $DMF:THF:H_2O$, 60 °C, 16 h	88	3;7A
9-BBN derivatives of L-aspartic acid	Halogenated pyridine	$Pd(PPh_3)_4$, K_3PO_4 (aq.), THF, 50 °C, 2 h	89	3;7B
Alkyl organoboron reagents	Aromatic esters	$Ni(COD)_2$, dcype, CsF, toluene, 150 °C	91	3; 8A
Alkyl organoboron reagents	Aroyl fluorides	$Ni(COD)_2$, dppe, CsF, toluene/hexane, 140 °C	92	3;8B
Potassium alkyltrifluoroborates	Aryl halides/triflates and vinyl triflates	$PdCl_2(dppf) \cdot CH_2Cl$, Cs_2CO_3, $THF:H_2O$, reflux, 6-72 h	44,94	4;9B
Tertiary trifluoroborate salts	Aryl and heteroaryl chlorides and bromides	CatacXium-A-Pd G3, Cs_2CO_3, tol/water, 90 °C, 18 h	99	4;9C
Secondary alkyl β-trifluoroboratoketones and -esters	Aryl Bromides	$Ir[dFCF_3ppy]_2(bpy)PF_6$, $NiCl_2 \cdot dme$, dtbbpy, Cs_2CO_3, 2,6-lutidine, 1,4-dioxane, hv	10	4;10A
α-alkoxyalkyl- and α-acyloxyalkyltrifluoroborates	Aryl bromides	$Ir[dFCF_3ppy]_2(bpy)PF_6$, $Ni(COD)_2$, dtbbpy, K_2HPO_4, dioxane, hv	101	4;10B
Tertiary organotrifluoroborates reagents	Aryl bromides	$Ir[dFCF_3ppy]_2(bpy)PF_6$, $Ni(TMHD)_2$ or $Ni(dtbbpy)(H_2O)_4Cl_2$, K_2HPO_4 or Na_2CO_3, no additive or $ZnBr_2$, dioxane/DMA or DMA, hv, 12–72 h	102	4;10C
Trialkylboranes	Aryl bromides	$PdCl_2(dppf)$, THF, reflux, 2–6 h	105,106	5;11B
NHC-boranes complexes	Aryl halides and triflates	[Pd], Ligand, tol-H_2O or THF-H_2O, heat or microwave	107	5;11C
Trialkyl- and triaryl-boranes (generated *in situ*)	Alkenyl and aryl halides	$Pd(OAc)_2$, n-$BuAd_2P$ or RuPhos, K_3PO_4, tol-H_2O, 100 °C	108	5;11D
n-alkylboronic acids	Alkenyl and aryl halides or triflates	$PdCl_2(dppf)$, K_2CO_3, Ag_2O, THF, 80 °C, 6–10 h	112	6;12A
n-alkylboronic acids	Alkenyl halides	$PdCl(C_3H_5)dppb$, Cs_2CO_3, toluene or xylene, 100–130 °C, 20 h	113	6;12B
Primary and secondary alkylboronic acids	2-bromoalken-3-ol derivatives	$Pd(OAc)_2$, LB-Phos.HBF_4, K_2CO_3, toluene, 110 °C, 3–27 h	114	6;12C

Table 1. Cont.

Boron Reagent	Substrate	Reaction Conditions (General)	Reference	Section and Scheme
Cyclic secondary alkylboronic acids	di-ortho-substituted arylhalides	Pd(OAc)$_2$, AntPhos, K$_3$PO$_4$, toluene, 110 °C, 12–24 h	115	6;12D
Acyclic secondary alkylboronic acids	Aryl and alkenyl triflates	[Pd(cinnamyl)Cl]$_2$, Ligand, K$_3$PO$_4$·H$_2$O, toluene, 110 °C, 12 h	116	6;12E
Boronic esters	Aryl methyl ethers bearing ortho-carbonyls	RuH$_2$(CO)(PPh$_3$)$_3$, toluene, 110 °C	117	7;13A
MIDA boronates	Aryl and heteroarylbromides	PdCl$_2$(dppf)·CH$_2$Cl$_2$, Cs$_2$CO$_3$, THF:H$_2$O, 80 °C, 24–48 h	43	7;14
Alkyl iodide and 9-MeO-9BBN (tBuLi for *in situ* generation)	Alkenyl bromide	Pd(OAc)$_2$, Aphos-Y	126	8;15B
OBBD derivatives	Aryl Bromide	Pd(dtbpf)Cl$_2$, Et$_3$N or K$_3$PO$_4$, TPGS-750-M/H$_2$O 45 °C, Ar, 16–21 h	127	8;15D

3. First Reports of B-alkyl SMC and Methods Employing 9-BBN Derivatives

The alkylboron cross-coupling was disclosed in 1986 by Suzuki and Miyaura using B-alkyl-9-BBN **2** or trialkylboranes (R$_3$B) in the presence of PdCl$_2$(dppf) and a base (sodium hydroxide or methoxide) (Scheme 3A). The reaction proceeded readily providing alkylated arenes **3** and alkenes in excellent yields of 75%–98%. On the other hand, no coupling was observed when sec-butylboranes were used [76]. In 1989, the same group revealed the reactivity of different alkyl boranes **5** in B-alkyl SMC (Scheme 3B). Pinacolborane **10** was almost unreactive (1% yield), while 9-BBN derivatives **7** showed the highest efficiencies (e.g., 99%). Thus, functionalized alkenes, arenes and cycloalkenes were synthesized via a hydroboration-coupling sequence of 9-BBN derivatives with haloalkenes or haloarenes **4** (inter- and intramolecular). Good yields of geometrically pure alkenes and arenes were afforded from the performed reactions with a variety of functionalities on either coupling partner. The reaction could also be carried out using K$_2$CO$_3$ instead of NaOH with base-sensitive compounds [77–79].

In 2004, the group of Buchwald reported the design of a new ligand with tuned steric and electronic properties. The phosphane ligand incorporated two methoxy groups on one of phenyls (**L2**, Scheme 3C). The oxygen lone pairs increase the electron density on the biaryl and participate in stabilizing the Pd complex. Simultaneously, the MeO groups increase the steric bulk and prevent cyclometalaton. This as-designed ligand aimed to serve as a universal catalyst for cross-coupling and C–H activation reactions. It was later commercialized under the name of SPhos, and became a basic ligand in today's catalysis toolbox. The ligand demonstrated a wide scope and stability with aryl boronic acids. It was also efficient for coupling of B-alkyl-9-BBN derivatives **14** (and boronic acids) using K$_3$PO$_4$·H$_2$O as an essential base (vs. lower conversions with anhydrous bases) (Scheme 3C). The scope involved challenging aryl halides as 3-dimethylamino-2-bromoanisole and aryl chlorides [80].

In 2013, Wu et al. developed a SMC between B-benzyl-9-BBN **18** and chloroenynes **16** and **17** to synthesize a vast array of 1,5-diphenylpent-3-en-1-yne derivatives **19** and **20** in good yields and full control on the E/Z selectivity using Pd(PPh$_3$)$_4$ and Cs$_2$CO$_3$ in pure water (Scheme 4) [81]. The conditions tolerated substrates bearing several electron-donating and withdrawing groups. It is worth remarking that these derivatives are known for their anti-inflammatory activity and can be isolated from plants, but only in minor quantities.

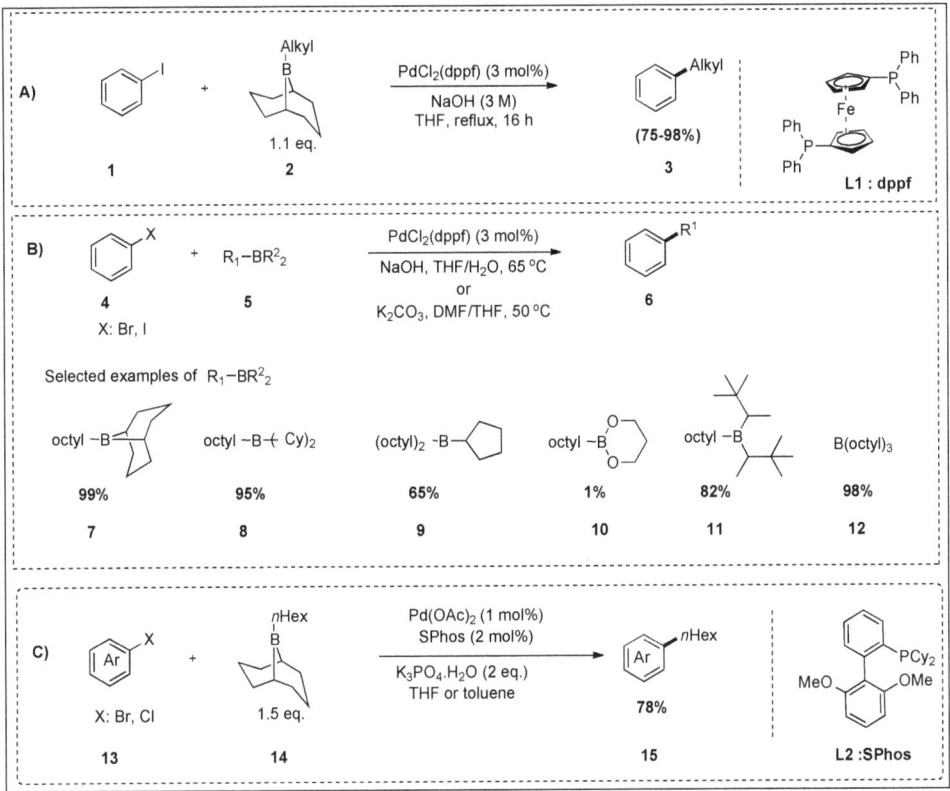

Scheme 3. First reports of B–alkyl Suzuki–Miyaura cross-coupling (**A–C**) and the reactivity of alkylboranes (**C**).

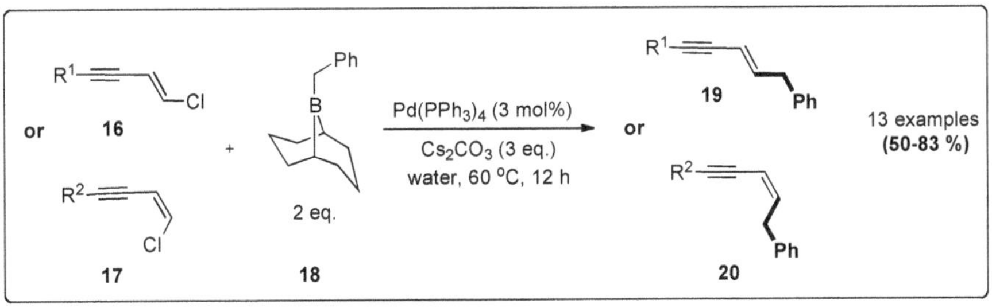

Scheme 4. B-alkyl SMC of chloroenynes.

C–O electrophiles represent attractive alternatives to halides. However, research on cross-couplings of aryl methyl ethers was delayed by the perception that they can be challenging coupling counterparts in comparison to other protected phenol electrophiles such as aryl pivalates, sulfonates and carbamates. Indeed, the activation energy for effecting C–OMe bond cleavage is significantly higher, with OMe being more difficult to separate from the group and more reluctant to oxidative addition. It is noteworthy that C–O electrophiles cross-couplings are predominantly conducted with nickel catalysis, as can be seen in Scheme 5, which depicts the work of the Rueping group in this regard. This demonstrates the higher activity of Ni with such challenging substrates [82–84].

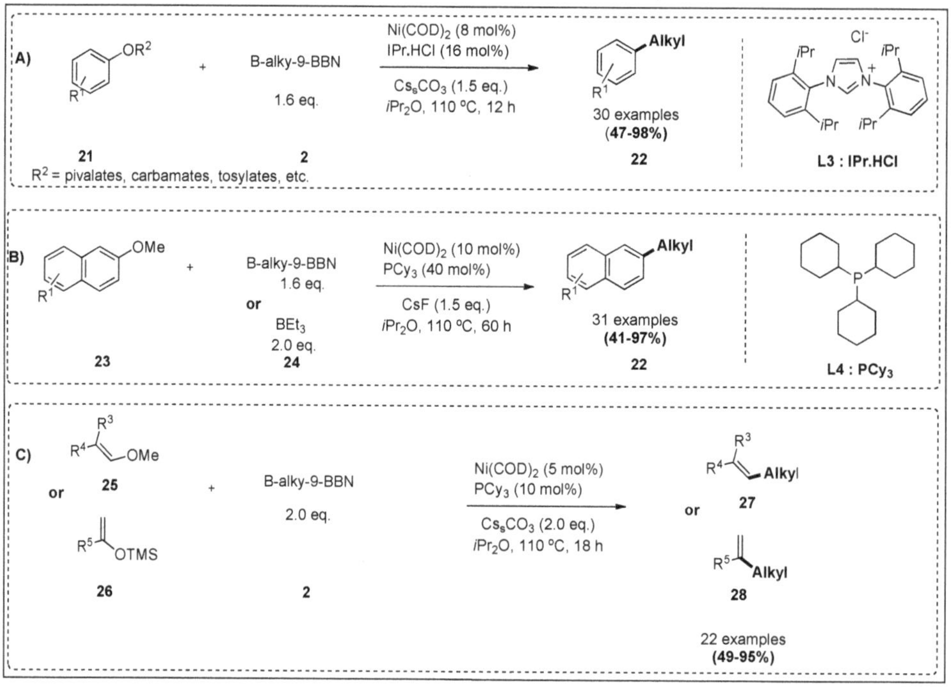

Scheme 5. Ni-catalyzed alkylation of C_{Ar}–O electrophiles (including aromatic methyl ethers) (**A**,**B**) and methyl enol ethers (**C**).

In 2016, Rueping et al. utilized the B-alkyl-9-BBN **2** to report an efficient nickel-catalyzed alkylation of C_{Ar}–O electrophiles **21** (pivalates, carbonates, carbamates, sulphamates and tosylates). The optimal conditions involved Ni(COD)$_2$, a IPr·HCl ligand and Cs$_2$CO$_3$ in diisopropyl ether (Scheme 5A). This new protocol was tolerant to numerous synthetically important functional groups of phenol pivalates and alkylboranes circumventing the restriction of β-hydride elimination [85]. Soon after, the same group described the use of the first alkylation of polycyclic aromatic methyl ethers **23** and methyl enol ethers **25** and **26** (Scheme 5B,C), which involves the cleavage of the highly inert C(sp^2)–OMe bonds, using alkylboron reagents with broad functional group tolerance. As expected, the choice of the base and the ligand is critical in C–O bond-cleaving reactions. Thus, the conditions described for C_{Ar}–O electrophiles were not successful, and the optimal conditions necessitated the replacement of the IPr·HCl ligand with PCy$_3$. Cs$_2$CO$_3$ was mostly used in couplings of alkenyl ethers, while both CsF and Cs$_2$CO$_3$ could be used in the case of aromatic methyl ethers. The reaction performed better with a Ni/L ratio of 1:4 instead of 1:2, and a prolonged reaction time of 60 h instead of 12 h. The optimal conditions for these novel transformations are summarized in Scheme 5 [86].

In 2018, Zhang et al. reported a hydroboration/Pd-catalyzed migrative SMC of 1,3-dienes **30** with electron-deficient aryl halides **29** (Scheme 6) with a wide scope (>40 examples). This method allows the use of primary homoallylic alkylboranes in the direct synthesis of branched allylarenes. The selectivity of the branched versus linear coupling was found to be tuned by the choice of the ligand. The branch-selective coupling was found to be favored by the more electron-rich bidentate ligand with a larger ligand–metal–ligand (bite) angle (i.e., **L5**: dppb). Their report involved preliminary mechanistic studies, showing a palladium migration in the formation of allyl palladium species. The migration proceeded via a sequence of β-hydride elimination and an alkene reinsertion partially involving an alkene dissociation/association process (Scheme 6) [87].

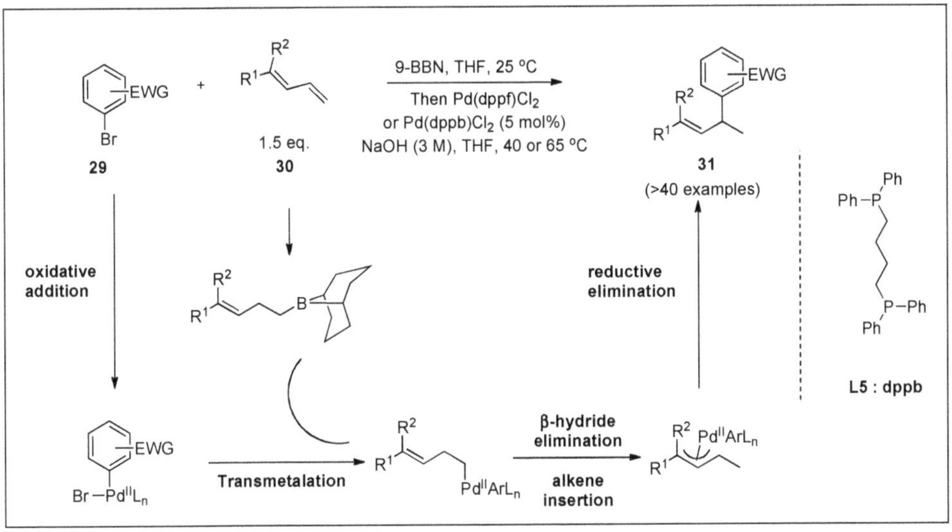

Scheme 6. Hydroboration/Pd-catalyzed migrative SMC of 1,3-dienes aryl halides.

Very recently, Newhouse et al. described the use of β-triflyl enones **32** as efficient coupling partners in a mild B–alkyl SMC (Pd(dppf)Cl$_2$ (2.5 mol%), Cs$_2$CO$_3$ (2 eq.)), and tolerant of sensitive functional groups (Scheme 7A). The more stable triflate to light and chromatography, in comparison to halogenated analogs, were used to establish challenging cyclic α,β-disubstituted enones **33** with good to excellent yields (10 examples) [88]. In parallel, Usuki et al. reported an SMC between halogenated pyridines **34** and a borated L-aspartic acid derivative (9-BBN) **35** using Pd(PPh$_3$)$_4$ (5 mol%) and K$_3$PO$_4$(aq.) in THF (Scheme 7B). The experimental yield gave insight on the reactivity order of halogen substituents and position, which was found to be as follows: Br > I >> Cl and C3 > C2, C4 [89].

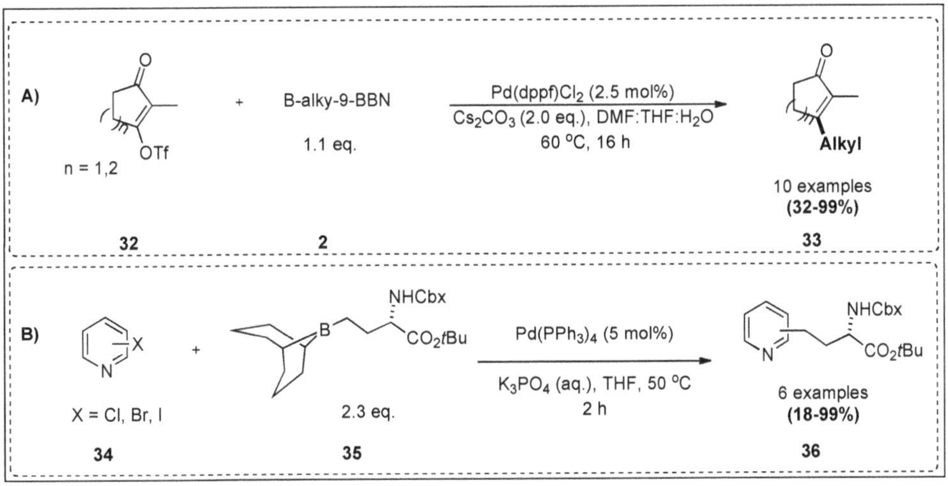

Scheme 7. Latest reports of SMCs using 9BBN (**A** and **B**).

Although decarbonylative and acyl cross-coupling reactions are not covered in this review [48–52,90], it is worth mentioning two very recent novel reports from the groups of Rueping and Nishihara. Rueping et al. (Scheme 8A) described an elegant ligand-controlled and site-selective nickel catalyzed SMC with aromatic esters **37** and alkyl organoboron reagents (majorly 9-BBN **2** and 6 examples with triethylboron). Ester substrates **37** were transformed into alkylated arenes **38** and ketone products **39** simply by switching the ligand from bidentate phosphine (**L6**: dcype) to monodentate phosphine (P(nBu)$_3$ or PCy$_3$). The regioselectivity was rationalized by DFT studies and the reported method has shown broad tolerance to functional groups and a wide substrate scope. The reaction was tested successfully on a large scale (1 g) using a cheaper NiCl$_2$ catalyst [91]. The group of Nishihara reported an elegant nickel-catalyzed decarbonylative C–F bond alkylation of aroyl fluorides **40**; the conditions are depicted in Scheme 8B [92].

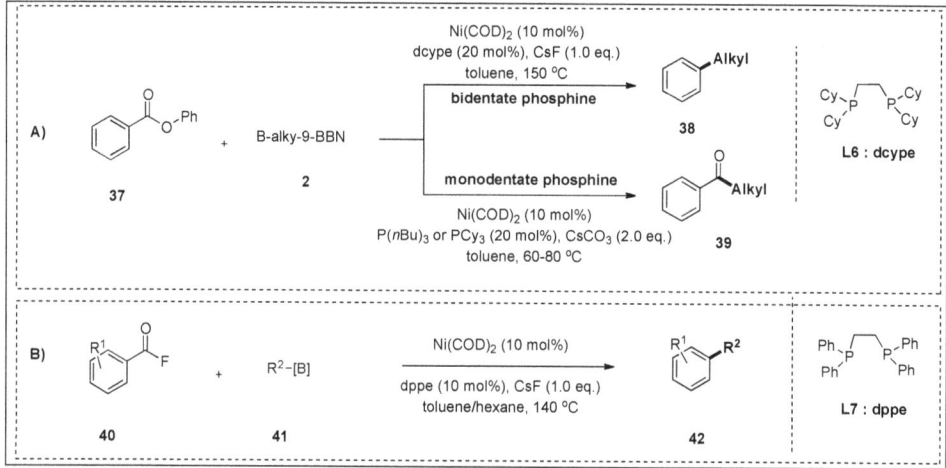

Scheme 8. Novel decarbonylative cross-coupling reactions with alkylboranes (**A** and **B**).

4. Organotrifluoroborates in sp^3–sp^2 SMCs

The tetracoordinate nature of the boron in organotrifluoroborates fortified by strong boron–fluorine bonds has been found to inhibit the undesirable reactions typical of trivalent organoborons. All of these complexes are crystalline solids and stable in water and under air; thus they can be stored on the shelf indefinitely. Besides, the manipulation of remote functional groups within the organotrifluoroborates is feasible while retaining the valuable C–B bond. Borates (RBF$_3$K) **45** can be easily prepared on a large scale by the addition of inexpensive fluoride source (KHF$_2$) **44** to a variety of organoboron intermediates **43**, such as boronic acids/esters, organodihaloboranes and organodiaminoboranes (Scheme 9A) [93].

Molander and coworkers were the first to use potassium alkyltrifluoroborates **45** as coupling partners with aryl halides/triflates and vinyl triflates **46/47** using PdCl$_2$(dppf)·CH$_2$Cl$_2$ as the catalyst in THF-H$_2$O and Cs$_2$CO$_3$ as the base (Scheme 9B). Two successive reports in 2001 and 2003 studied the scope of this B-alkyl SMC, reporting more than 50 examples with acceptable to very good yields, hence revealing a potential general method to a wide range of functionalities [44,94]. Later, the same group used microscale parallel experimentation to describe the first comprehensive study of the coupling of secondary alkylborons (organotrifluoroborates) and aryl chlorides (and bromides), elaborating different catalytic systems for this purpose. Their results demonstrated a ligand-dependent β-hydride elimination/reinsertion mechanism in the cross-couplings of hindered partners, which can result in isomeric products of coupled products [34]. The use of trifluoroborates in SMC was validated by numerous publications that appeared thereafter and was reviewed many times by different research groups, as the one by Molander in 2015 [79,95–98].

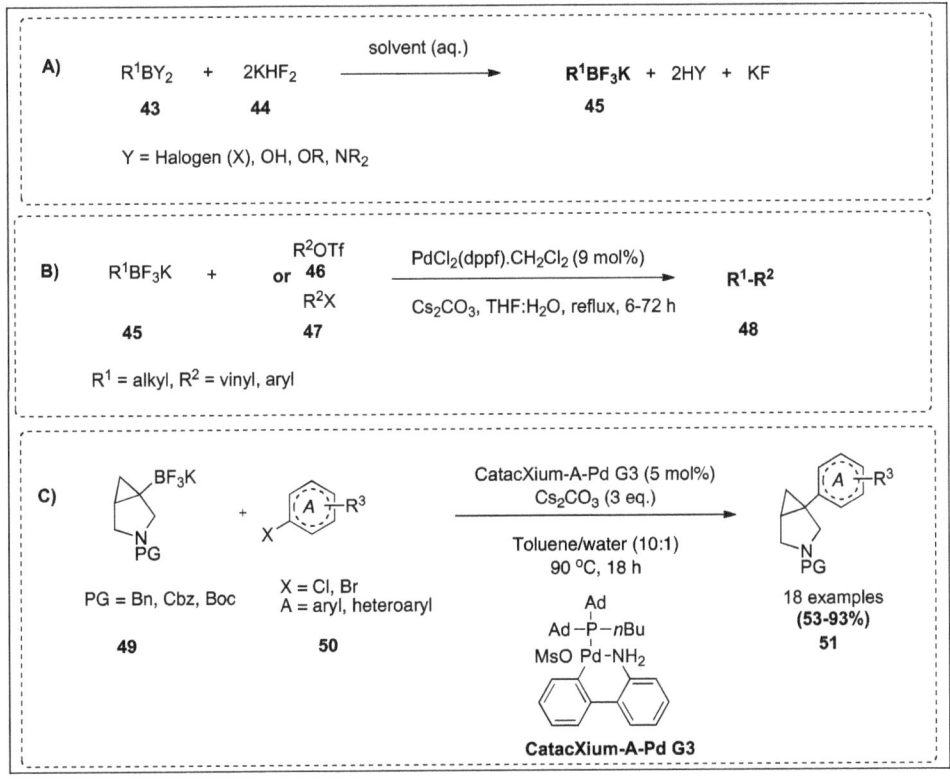

Scheme 9. Alkyltrifluoroborates salts: General synthesis and first report in sp³–sp² SMC (**A** and **B**); Pd-catalyzed SMC report of Harris et al. (**C**).

Harris et al. recently reported a Pd-catalyzed SMC reaction with tertiary trifluoroborate salts **49** to synthesize 1-heteroaryl-3-azabicyclo[3.1.0]hexanes **51**, an interesting scaffold in medicinal studies with limited synthetic approaches. The SMC protocol was compatible with a range of aryl and heteroaryl chlorides and bromides **50** (Scheme 9C) [99]. The optimized conditions involved CatacXium-A-Pd-G3, Cs_2CO_3 in toluene/water and were applied in synthesis of 18 examples with good to excellent yields.

The group of Molander, after their review [98], has extended the scope of sp²–sp³ cross-couplings to fluoroborates that show recalcitrance to Pd-catalyzed classical couplings via dual catalysis (Scheme 10). The first comprised the coupling of aryl bromides **53** to secondary alkyl β-trifluoroboratoketones and -esters **52** using Ir-based photoredox/nickel dual catalysis (Scheme 10A). This dual catalysis relies on a single-electron transmetalation and provides a complementary toolbox to the classical couplings that are based on two-electron processes. The oxidative fragmentation in the dual catalysis activates the organometallic reagent into its corresponding alkyl radical, which is then readily intercepted by the nickel catalyst mediating the formation of the C–C bond formation with the aryl halide partner. Their optimized conditions consisted of a catalytic system of Ir[dFCF₃ppy]₂(bpy)PF₆ photocatalyst (2.5 mol%), NiCl₂·dme (2.5 mol%), dtbbpy (2.5 mol%), Cs_2CO_3 (0.5 eq.) and 2,6-lutidine (0.5 eq.) in 1,4-dioxane, tolerating various functionalities in addition to sterically and electronically diverse coupling partners (Scheme 10A) [100]. The second report described a photoredox/nickel dual catalysis alternative approach to the protecting-group-independent cross-coupling of α-alkoxyalkyl- and α-acyloxyalkyltrifluoroborates **55** with aryl (and heteroaryl) bromides **53**, which can also be achieved by palladium catalysis. This method was compatible with various functional groups and N,N-diisopropylcarbamoyl, pivaloyl and benzyl protecting groups (Scheme 10B) [101]. Their third dual catalysis report (Scheme 10C) contributed to the construction of sterically demanding quaternary centers **58**, an area that is not yet comprehensive and suffers from the absence of general methodologies and the copious limitations of the currently used metal-catalyzed methods. Various tertiary organotrifluoroborates reagents **57** were coupled using different conditions and light intensities, which were found to be crucial depending on the nature of the substituents (e.g., bridged versus acyclic). The scope of the coupled aryl bromides **53** in this method was limited to electron-poor and electron-neutral systems [102].

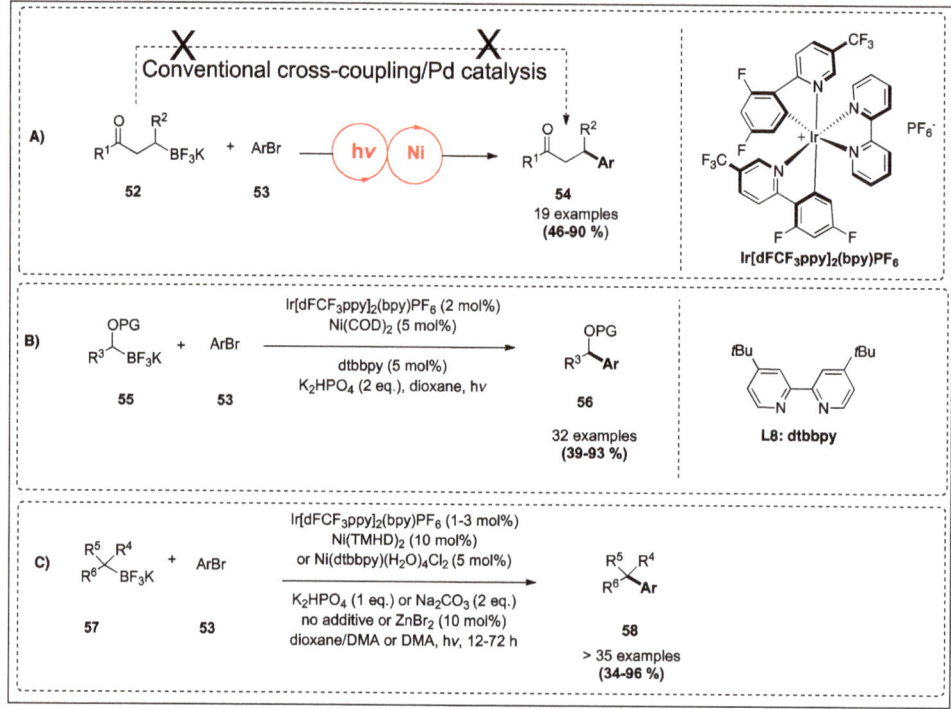

Scheme 10. Photoredox/metal dual catalysis of organotrifluoroborates by the Molander group (A–C).

5. Other Alkylboranes in sp^3–sp^2 SMCs

Tri-*n*-alkylboranes (R$_3$B) can be easily prepared by the reaction of Grignard reagents with boron trifluoride etherate (Scheme 11A) [103]. The use of this class of boranes in B–alkyl SMC was sporadically reported in the literature, probably due to their flammable nature and sensitivity to oxygen, as well as the inefficiency of the transfer of all three alkyl groups from the boron center [104]. In 2009, Wang et al. published optimization studies that presented efficient and chemoselective Pd-catalyzed direct SMCs of trialkylboranes **60** with bromoarenes **59** in the presence of unmasked acidic or basic functions using the weak base Cs$_2$CO$_3$ under mild non-aqueous conditions (Scheme 11B). The conditions tolerated carbonyl reagents, chlorinated derivatives, nitriles and unprotected and base-labile Piv- and TBS-protected phenols with more than 30 examples incorporating primary alkyls, and especially lower *n*-alkyls such as ethyl groups [105,106].

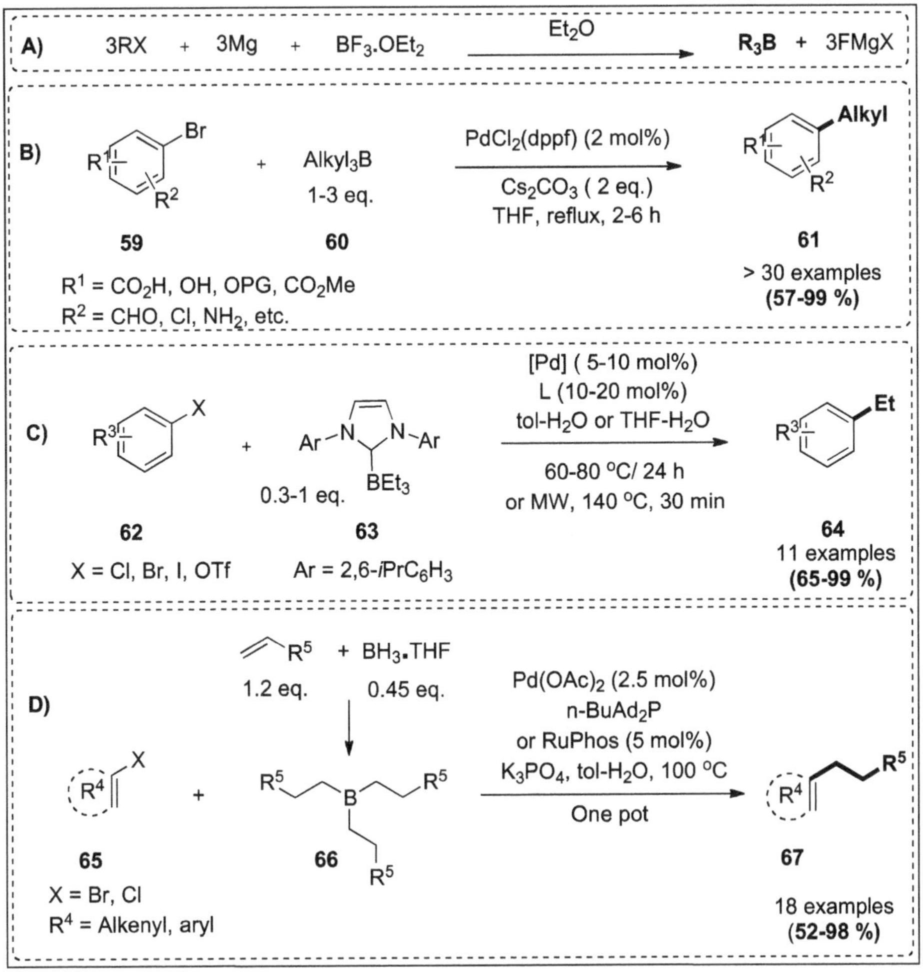

Scheme 11. Synthesis of alkylboranes (**A** and **D**) and their uses as coupling partners in sp^3–sp^2 SMCs (**A**–**D**).

Lacôte et al. developed the efficient transfer of all three groups of trialkyl- and triaryl-boranes (0.3–1 eq. instead of 1–3 eq.) in SMC in good yields under base-free conditions, achieving the activation by using N-heterocyclic carbenes (i.e., **63** in Scheme 11C). The C(sp^2)-C(sp^3) scope involved the NHC–borane complexes **63** with aryl chlorides, bromides, iodides and triflates **62** in 11 examples (65%–99%) using PdCl$_2$(dppf) or Pd(OAc)$_2$ with a ligand (XPhos or RuPhos) under microwave irradiation or classical heating [107]. In 2015, Li et al. described a general, atom-economic methodology that uses peralkyl and peraryl groups of unactivated symmetrical triaryl- and trialkyl-boranes **66** in SMC (Scheme 11D). The hydroboration of terminal alkenes was carried out *in situ*, and the corresponding trialkylboranes **66** were coupled with alkenyl and aryl halides **65** in a one-pot fashion. The method was compatible with a variety of functional groups and heterocycles [108].

6. Alkylboronic Acids in sp^3–sp^2 SMCs

Alkylboronic acids (R(BOH)$_2$), like their aryl analogs, exist in equilibrium with their trimeric cyclic anhydrides—boroxines, which also proved to be efficient coupling partners in SMCs [109]. Thus, the determination of the concentration of boroxine vs. boronic acid in the catalytic reaction can be difficult, requiring the employment of excess boronic acid to ensure the completion of the reaction [110]. Gibbs et al. were among the first to use alkylboronic acids as coupling partners with alkenyl triflates in 1995 [111]. The group of Falck widened the scope by reporting an efficient Ag(I)-promoted SMC of n-alkylboronic acids 68 (Scheme 12A) [112].

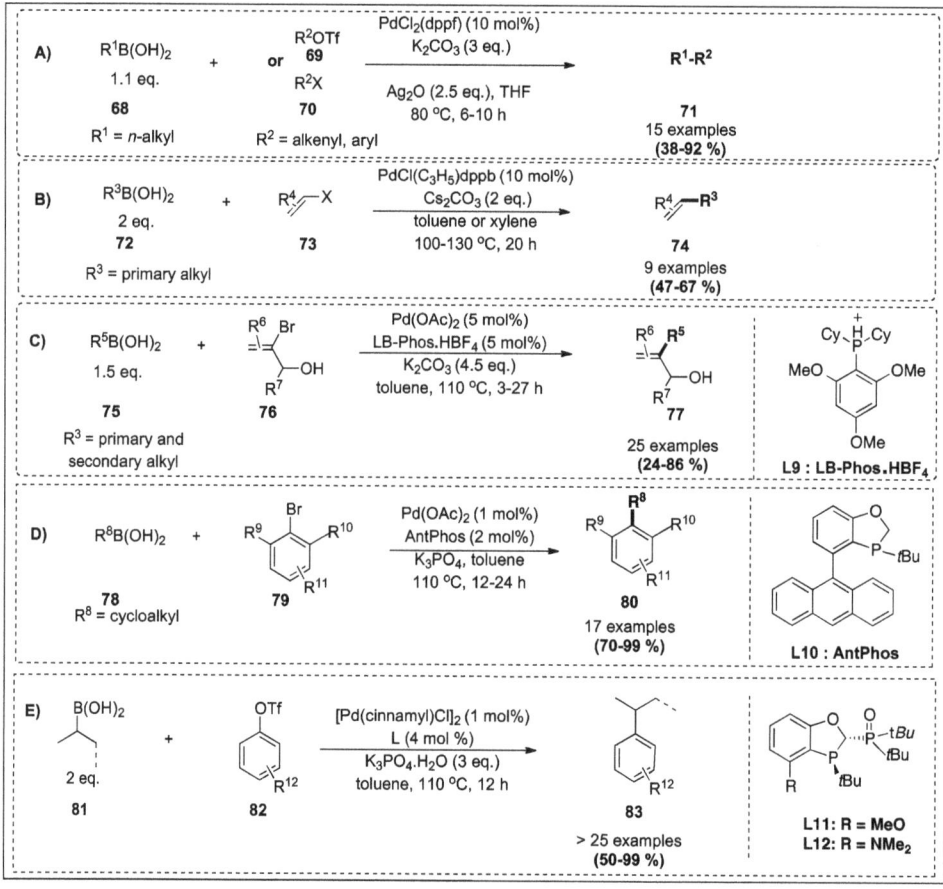

Scheme 12. Alkylboronic acids as coupling partners in sp^3–sp^2 SMCs (A–E).

The progress of utilizing alkylboronic acids was reviewed in 2008 [110]. Next, the SMC of primary alkylboronic acids 72 with alkenyl halides 73 was reported using air-stable catalyst PdCl(C$_3$H$_5$)(dppb) and Cs$_2$CO$_3$, and toluene or xylene as solvents (Scheme 12B) [113]. In 2012, Ma et al. used Pd(OAc)$_2$ with K$_2$CO$_3$ and an air-stable monophosphine HBF$_4$ salt (L9: LB-Phos.HBF$_4$) as an efficient ligand to couple primary and secondary alkylboronic acids 75 with 2-bromoalken-3-ol derivatives 76 (Scheme 12C) [114]. In 2014, Tang et al. revealed a sterically demanding aryl–alkyl SMC between di-ortho-substituted arylhalides 79 and (secondary) cycloalkylboronic acids 78 using a highly reactive Pd-AntPhos catalyst that allowed to reduce the β-hydride elimination (Scheme 12D). The method comprised a scope of sterically hindered substituted aryl compounds, including highly substituted benzene, naphthalene and anthracene derivatives [115]. The same group described the cross-coupling between aryl/alkenyl triflates 82 and acyclic secondary alkylboronic acids 81 in good to excellent yields (Scheme 12E). The employment of sterically bulky P,P=O ligands (L11/12) was found to be critical to achieve the chemoselectivity by inhibiting the isomerization of the secondary alkyl coupling partner (e.g., iPr vs. nPr) and to obtain high yields [116].

7. Boronic Esters and MIDA Boronates in sp^3–sp^2 SMCs

Prior to the work of Rueping on more general cross-coupling methods of challenging C–O electrophiles with organoboron reagents, a robust Ru-catalyzed SMC of aryl methyl ethers 84 with boronic esters 85 was elegantly revealed by chelation assistance (Scheme 13A) [84,117]. Aromatic ketones 84 where the carbonyl is located in an ortho position were reported to assist in the cleavage of C–OMe bonds. Neopentyl boronates 85 were the most reactive among all the tested boronic esters. The conditions were employed to couple aryl, alkenyl and even alkyl boronates with the same efficiency by using a RuH$_2$(CO)(PPh$_3$)$_3$ catalytic system. The C–OMe bond-cleavage was facilitated by the coordination of the carbonyl group to the Ru center, in an analogous mechanistic scenario to C–H activation (Scheme 13B). The suggested chelation-assisted mechanism was later supported by the isolation of the oxidative addition complex of an aryl C–O bond using low-valent Ru complexes 91 (Scheme 13C) [84,118,119]. The C–O bond-cleavage occurred at high temperatures (thermodynamic control) as compared to the C–H functionalization that rapidly took place at room temperature (Scheme 13C). The Ru-catalyzed SMC of aryl methyl ethers remained restricted to the presence of an ortho directing group to the reactive site [84,118,119]. The reported more general Ni-catalyzed coupling version of aryl methyl ether without directing group involved aryl boranes, and did not involve a scope of alkyl boranes [84,117–120].

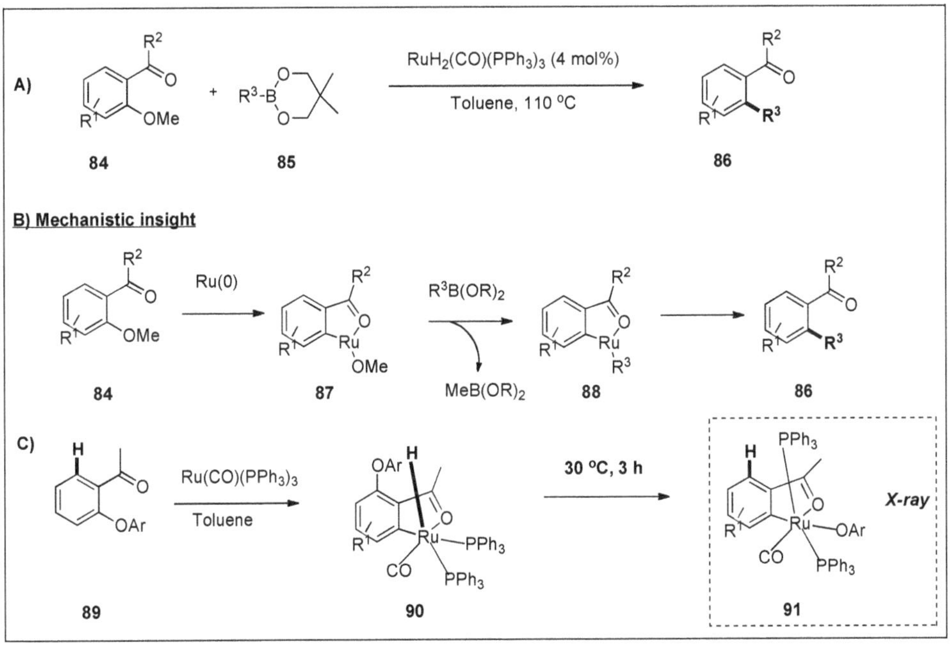

Scheme 13. Chelation-assisted Ru-catalyzed sp^3–sp^2 SMCs of C–OMe electrophiles (**A**) and mechanistic insight (**B,C**).

Inspired by the pioneering work of Wrackmeyer on protected boronic acids by iminodiacetic acids [121], the groups of Burke, Yudin and others developed the use of N-methyliminodiacetic acid (MIDA) boronates 92 in direct and iterative SMC reactions [122–124]. In addition to stability and compatibility with chromatography, the advantage of MIDA boronates is their mild hydrolysis to liberate the corresponding boronic acids compared to the harsh conditions needed in the case of sterically bulky boronic esters. This class found various applications in synthesis, and the efficient iterative assembly of the MIDA building blocks was recently reviewed in 2015 [122]. A direct SMC between MIDA boronates **92** and aryl and heteroaryl bromides **93** is presented in Scheme 14 [43].

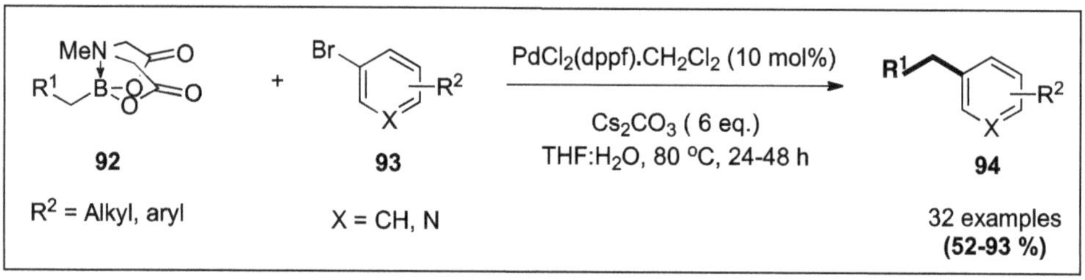

Scheme 14. sp^3–sp^2 SMCs using N-methyliminodiacetic acid (MIDA) boronates.

8. B–Alkyl SMCs Using BBN Variants (9-MeO-9-BBN and OBBD Derivatives)

The basic set-up of the SMC has essentially stayed similar for decades. However, the '9-MeO-9-BBN variant' is one of the alternative formats for this transformation that has permitted advanced applications of the sp^3–sp^2 coupling process (Scheme 15A,B). This method is distinguished by the absence of the essential base that acts as a promoter in the classical SMC version. Rather, the R–M (sp^3, sp^2, or sp) is first intercepted with 9-MeO-9-BBN, resulting in the corresponding borinate complex **97**, which then passes the R-group onto an organopalladium complex generated *in situ* as the electrophilic partner (Scheme 15A). The 9-MeO-9-BBN variant was reviewed by Seidel and Fürstner in 2011 [125]. In 2013, Dai et al. reported a 9-MeO-9-BBN variant methodology, depicted in Scheme 15B, using Pd(OAc)$_2$ and a hemilabile P,O-ligand, Aphos-Y **L13** under mild reaction conditions ($K_3PO_4·3H_2O$, THF/H_2O, rt) coupling the alkyl iodide **99** and the alkenyl bromide **100**. This new process serves as an improvement of the Johnson protocol, which generally employs two ligands (dppf and Ph$_3$As) and two organic solvents (THF and DMF) in the SMC step in the total synthesis of structurally complex natural products, by using one ligand (**L13**, Aphos-Y) and one organic solvent (THF) [126].

OBBD (B-alkyl-9-oxa-10-borabicyclo[3.3.2]decane) derivatives **104/105** represent another variant of 9-BBN (Scheme 15C,D). OBBD reagents **104/105** were used successfully to perform B-Alkyl SMC under mild aqueous micellar catalysis conditions. The straightforward preparation of OBBD **104/105** is shown in Scheme 15C.

OBBD derivatives showed similar reactivity to 9-BBN reagents in SMCs, with the advantage of increased stability and isolable nature. The optimized SMC conditions (Scheme 15D) comprised dtbpf **L14** as the supporting ligand, which allows the reaction to be run at a catalyst loading as low as 0.25 mol% (i.e., 2500 ppm). The optimization was carried out in aqueous surfactant media, with TPGS-750-M as the preferred amphiphile and Et$_3$N or K$_3$PO$_4$ as the base. The substrate scope **108** was shown by more than 34 examples with good to excellent yields (56%–100%). Lower yields were observed with steric hindrance next to the boronate group, and the conditions were limited on secondary OBBD reagents (even upon using 9-BBN derivatives instead). The synthetic utility of this methodology was demonstrated by a four-step one-pot synthesis and a successful recycling of the reaction medium [127].

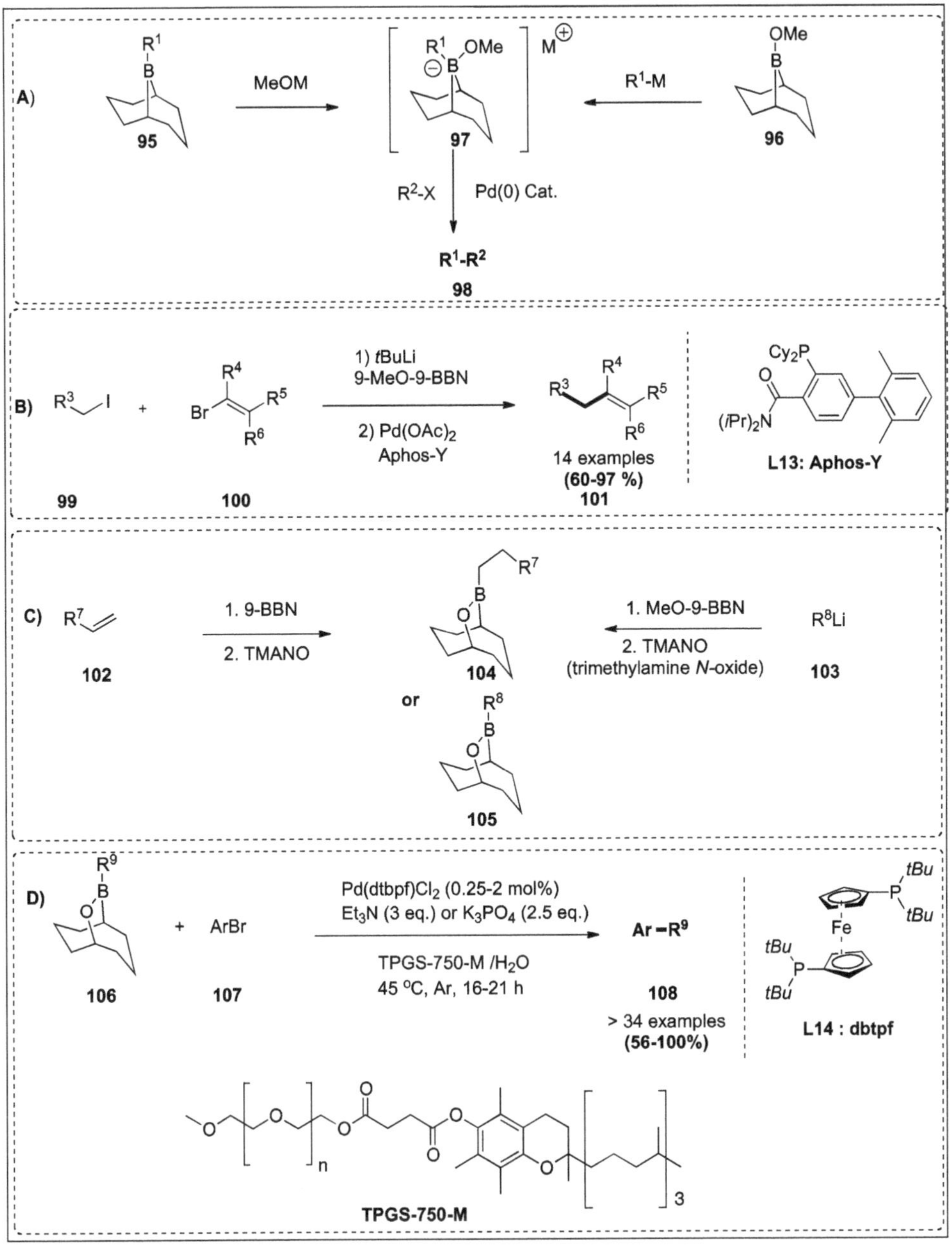

Scheme 15. B–alkyl SMCs using BBN variants (9-MeO-9-BBN (**A**,**B**)) and OBBD derivatives (**C**,**D**)).

9. Selected Examples of Applications of SMCs and B–alkyl SMC in the Synthesis of Target Molecules

It is rare nowadays to find a total synthesis that does not involve at least a cross-coupling reaction, and in particular, a Suzuki–Miyaura reagent [6]. The use of SMC in total synthesis has been extensively reviewed by Heravi et al. [128,129].

B–alkyl SMC, in particular, was likewise applied in the synthesis of beneficial products [130–132]. Two examples are shown in Scheme 16: **Cytochalasin Z₈** and **Ieodomycin D**, which belong to the family of secondary fungal metabolite with a wide range of biological activities that target cytoskeletal processes [133–135]. Scheme 16 also includes examples of complex molecules that were achieved by synthetic routes involving SMCs with $C(sp^2)$–B reagents; namely **Michellamine** (an anti-HIV viral replication receptor) and **(-)-steganone** (an antileukemic lignan precursor) [136,137].

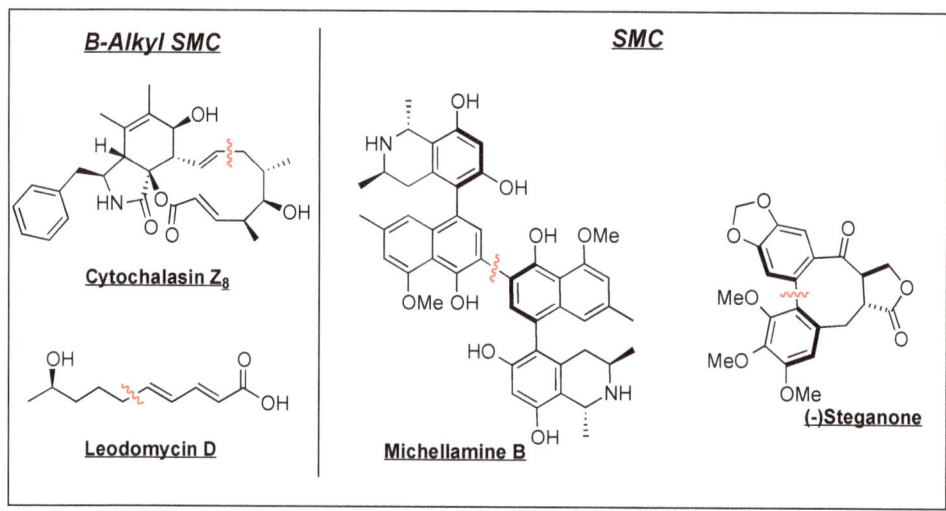

Scheme 16. Examples of drugs and active molecules whose total synthesis involved SMC.

10. Conclusion

The present review focused on the use of $C(sp^3)$–organoboranes as cross-coupling partners in metal-catalyzed $C(sp^3)$–$C(sp^2)$ cross-couplings, such as B–alkyl Suzuki–Miyaura reactions. Indeed, metal-catalyzed cross-coupling reactions have become mature tools in organic synthesis. Nevertheless, $C(sp^3)$–C cross-couplings are far less reported than other C-C coupling reactions. Furthermore, this field is largely dominated by using organic halides or pseudohalides as coupling partners. C–O–Alkyl electrophiles remain an area of research that is attracting strong attention. Undoubtedly, the progress made in the syntheses of stable and isolable sp^3-boron reagents is impacting the development of $C(sp^3)$–$C(sp^2)$ cross-couplings of the Suzuki–Miyaura type. The attention given to dual and photocatalysis is also strongly contributing to the furnishing of a toolbox that can achieve active adducts, which impact all fields of research and industry and cannot be otherwise obtained.

Author Contributions: J.S., J.E.-M. and T.M.E.D. wrote the manuscript; I.K. wrote the mechanistic insight (Section 2) and proofread the whole manuscript; C.-S.L., A.K. and K.P. and proofread the manuscript and provided critical revision throughout the process. All authors have read and agreed to the published version of the manuscript.

Funding: This publication is based upon work supported by the Khalifa University of Science and Technology under Award No. RC2-2018-024".

Conflicts of Interest: "The authors declare no conflict of interest."

References

1. Frenking, G. Peculiar boron startles again. *Nature* **2015**, *522*, 297–298. [CrossRef]
2. Kumar, R.; Karkamkar, A.; Bowden, M.; Autrey, T. Solid-state hydrogen rich boron-nitrogen compounds for energy storage. *Chem. Soc. Rev.* **2019**, *48*, 5350–5380. [CrossRef]

3. William Lipscomb Biographical. Available online: https://www.nobelprize.org/prizes/chemistry/1976/lipscomb/biographical/ (accessed on 22 February 2020).
4. Brown, H.C. From Little Acorns to Tall Oaks: From Boranes through Organoboranes. Available online: https://www.nobelprize.org/prizes/chemistry/1979/summary/ (accessed on 22 February 2020).
5. Miyaura, N.; Suzuki, A. Palladium-Catalyzed Cross-Coupling Reactions of Organoboron Compounds. *Chem. Rev.* **1995**, *95*, 2457–2483. [CrossRef]
6. Maluenda, I.; Navarro, O. Recent developments in the Suzuki-Miyaura reaction: 2010–2014. *Molecules* **2015**, *20*, 7528–7557. [CrossRef] [PubMed]
7. Suzuki, A.; Yamamoto, Y. Cross-coupling Reactions of Organoboranes: An Easy Method for C–C Bonding. *Chem. Lett.* **2011**, *40*, 894–901. [CrossRef]
8. Defrancesco, H.; Dudley, J.; Coca, A. Boron Chemistry: An Overview. *ACS Symp. Ser.* **2016**, *1236*, 1–25.
9. Brown, H.C.; Cole, T.E. Organoboranes. 31 A Simple Preparation of Boronic Esters from Organolithium Reagents and Selected Trialkoxyboranes. *Organometallics* **1983**, *2*, 1316–1319. [CrossRef]
10. Fyfe, J.W.B.; Watson, A.J.B. Recent Developments in Organoboron Chemistry: Old Dogs, New Tricks. *Chem* **2017**, *3*, 31–55. [CrossRef]
11. Dimitrijević, E.; Taylor, M.S. Organoboron acids and their derivatives as catalysts for organic synthesis. *ACS Catal.* **2013**, *3*, 945–962. [CrossRef]
12. Akira Suzuki Nobel Lecture Nobel Prizes 2019. Available online: https://www.nobelprize.org/prizes/chemistry/2010/suzuki/lecture/ (accessed on 22 February 2020).
13. Das, B.C.; Thapa, P.; Karki, R.; Schinke, C.; Das, S.; Kambhampati, S.; Banerjee, S.K.; Van Veldhuizen, P.; Verma, A.; Weiss, L.M.; et al. Boron chemicals in diagnosis and therapeutics. *Future Med. Chem.* **2013**, *5*, 653–676. [CrossRef]
14. Klotz, J.H.; Moss, J.I.; Zhao, R.; Davis, L.R.; Patterson, R.S. Oral toxicity of boric acid and other boron compounds to immature cat fleas (Siphonaptera: Pulicidae). *J. Econ. Entomol.* **1994**, *87*, 1534–1536. [CrossRef] [PubMed]
15. Ban, X.; Jiang, W.; Sun, K.; Xie, X.; Peng, L.; Dong, H.; Sun, Y.; Huang, B.; Duan, L.; Qiu, Y. Bipolar host with multielectron transport benzimidazole units for low operating voltage and high power efficiency solution-processed phosphorescent OLEDs. *ACS Appl. Mater. Interfaces* **2015**, *7*, 7303–7314. [CrossRef] [PubMed]
16. Jäkle, F. Recent Advances in the Synthesis and Applications of Organoborane Polymers. In *Synthesis and Application of Organoboron Compounds*; Fernández, E., Whiting, A., Eds.; Springer International Publishing: Cham, Switzerland, 2015; pp. 297–325. ISBN 978-3-319-13053-8.
17. Matsumi, N.; Sugai, K.; Sakamoto, K.; Mizumo, T.; Ohno, H. Direct synthesis of poly(lithium organoborate)s and their ion conductive properties. *Macromolecules* **2005**, *38*, 4951–4954. [CrossRef]
18. Liu, L.; Corma, A. Metal Catalysts for Heterogeneous Catalysis: From Single Atoms to Nanoclusters and Nanoparticles. *Chem. Rev.* **2018**, *118*, 4981–5079. [CrossRef]
19. Toffoli, D.; Stredansky, M.; Feng, Z.; Balducci, G.; Furlan, S.; Stener, M.; Ustunel, H.; Cvetko, D.; Kladnik, G.; Morgante, A.; et al. Electronic properties of the boroxine–gold interface: Evidence of ultra-fast charge delocalization. *Chem. Sci.* **2017**, *8*, 3789–3798. [CrossRef]
20. Chen, C.C.; Fan, H.J.; Shaya, J.; Chang, Y.K.; Golovko, V.B.; Toulemonde, O.; Huang, C.H.; Song, Y.X.; Lu, C.S. Accelerated $ZnMoO_4$ photocatalytic degradation of pirimicarb under UV light mediated by peroxymonosulfate. *Appl. Organomet. Chem.* **2019**, *33*, 1–15.
21. Chardon, A.; Mohy El Dine, T.; Legay, R.; De Paolis, M.; Rouden, J.; Blanchet, J. Borinic Acid Catalysed Reduction of Tertiary Amides with Hydrosilanes: A Mild and Chemoselective Synthesis of Amines. *Chem. Eur. J.* **2017**, *23*, 2005–2009. [CrossRef]
22. Shaya, J.; Deschamps, M.A.; Michel, B.Y.; Burger, A. Air-Stable Pd Catalytic Systems for Sequential One-Pot Synthesis of Challenging Unsymmetrical Aminoaromatics. *J. Org. Chem.* **2016**, *81*, 7566–7573. [CrossRef]
23. Papageridis, K.N.; Siakavelas, G.; Charisiou, N.D.; Avraam, D.G.; Tzounis, L.; Kousi, K.; Goula, M.A. Comparative study of Ni, Co, Cu supported on γ-alumina catalysts for hydrogen production via the glycerol steam reforming reaction. *Fuel Process. Technol.* **2016**, *152*, 156–175. [CrossRef]
24. Polychronopoulou, K.; Costa, C.N.; Efstathiou, A.M. The steam reforming of phenol reaction over supported-Rh catalysts. *Appl. Catal. A Gen.* **2004**, *272*, 37–52. [CrossRef]

25. Charisiou, N.D.; Siakavelas, G.; Papageridis, K.N.; Baklavaridis, A.; Tzounis, L.; Polychronopoulou, K.; Goula, M.A. Hydrogen production via the glycerol steam reforming reaction over nickel supported on alumina and lanthana-alumina catalysts. *Int. J. Hydrog. Energy* **2017**, *42*, 13039–13060. [CrossRef]
26. Karamé, I.; Shaya, J.; Srour, H. *Carbon Dioxide Chemistry, Capture and Oil Recovery*; IntechOpen: London, UK, 2018; ISBN 9781789235753.
27. Mohy El Dine, T.; Evans, D.; Rouden, J.; Blanchet, J. Mild Formamide Synthesis through Borinic Acid Catalysed Transamidation. *Chem. Eur. J.* **2016**, *22*, 5894–5901. [CrossRef] [PubMed]
28. Chen, C.C.; Shaya, J.; Fan, H.J.; Chang, Y.K.; Chi, H.T.; Lu, C.S. Silver vanadium oxide materials: Controlled synthesis by hydrothermal method and efficient photocatalytic degradation of atrazine and CV dye. *Sep. Purif. Technol.* **2018**, *206*, 226–238. [CrossRef]
29. Holstein, P.M.; Dailler, D.; Vantourout, J.; Shaya, J.; Millet, A.; Baudoin, O. Synthesis of Strained γ-Lactams by Palladium(0)-Catalyzed C(sp3)-H Alkenylation and Application to Alkaloid Synthesis. *Angew. Chem. Int. Ed.* **2016**, *55*, 2805–2809. [CrossRef]
30. Karamé, I.; Zaher, S.; Eid, N.; Christ, L. New zinc/tetradentate N4 ligand complexes: Efficient catalysts for solvent-free preparation of cyclic carbonates by CO2/epoxide coupling. *Mol. Catal.* **2018**, *456*, 87–95. [CrossRef]
31. Ridgway, B.H.; Woerpel, K.A. Transmetalation of Alkylboranes to Palladium in the Suzuki Coupling Reaction Proceeds with Retention of Stereochemistry. *J. Org. Chem.* **1998**, *63*, 458–460. [CrossRef]
32. Johnson, C.R.; Braun, M.P. A Two-step, Three-Component Synthesis of PGE1: Utilization of a-Iodoenones in Pd(0)-Catalyzed Cross-Couplings of Organoboranest. *J. Am. Chem. Soc.* **1993**, *115*, 11014–11015. [CrossRef]
33. Chemler, S.R.; Trauner, D.; Danishefsky, S.J. The B -alkyl Suzuki–Miyaura cross-coupling reaction: Development, mechanistic study, and applications in natural product synthesis. *Angew. Chem. Int. Ed.* **2001**, *40*, 4544–4568. [CrossRef]
34. Dreher, S.D.; Dormer, P.G.; Sandrock, D.L.; Molander, G.A. Efficient cross-coupling of secondary alkyltrifluoroborates with aryl chlorides-reaction discovery using parallel microscale experimentation. *J. Am. Chem. Soc.* **2008**, *130*, 9257–9259. [CrossRef]
35. Molander, G.A.; Wisniewski, S.R. Stereospecific cross-coupling of secondary organotrifluoroborates: Potassium 1-(benzyloxy)alkyltrifluoroborates. *J. Am. Chem. Soc.* **2012**, *134*, 16856–16868. [CrossRef]
36. González-Bobes, F.; Fu, G.C. Amino alcohols as ligands for nickel-catalyzed Suzuki reactions of unactivated alkyl halides, including secondary alkyl chlorides, with arylboronic acids. *J. Am. Chem. Soc.* **2006**, *128*, 5360–5361. [CrossRef] [PubMed]
37. Lu, Z.; Wilsily, A.; Fu, G.C. Stereoconvergent amine-directed alkyl-alkyl Suzuki reactions of unactivated secondary alkyl chlorides. *J. Am. Chem. Soc.* **2011**, *133*, 8154–8157. [CrossRef] [PubMed]
38. Sun, H.-Y.; Hall, D.G. *At the Forefront of the Suzuki–Miyaura Reaction: Advances in Stereoselective Cross-Couplings BT—Synthesis and Application of Organoboron Compounds*; Fernández, E., Whiting, A., Eds.; Springer International Publishing: Cham, Switzerland, 2015; pp. 221–242. ISBN 978-3-319-13054-5.
39. Lennox, A.J.J.; Lloyd-Jones, G.C. Selection of boron reagents for Suzuki–Miyaura coupling. *Chem. Soc. Rev.* **2014**, *43*, 412–443. [CrossRef] [PubMed]
40. Xu, L.; Zhang, S.; Li, P. Boron-selective reactions as powerful tools for modular synthesis of diverse complex molecules. *Chem. Soc. Rev.* **2015**, *44*, 8848–8858. [CrossRef]
41. Suzuki, A. Organoboranes in organic syntheses including Suzuki coupling reaction. *Heterocycles* **2010**, *80*, 15–43. [CrossRef]
42. Nave, S.; Sonawane, R.P.; Elford, T.G.; Aggarwal, V.K. Protodeboronation of tertiary boronic esters: Asymmetric synthesis of tertiary alkyl stereogenic centers. *J. Am. Chem. Soc.* **2010**, *132*, 17096–17098. [CrossRef]
43. St. Denis, J.D.; Scully, C.C.G.; Lee, C.F.; Yudin, A.K. Development of the Direct Suzuki–Miyaura Cross-Coupling of Primary B -Alkyl MIDA-boronates and Aryl Bromides. *Org. Lett.* **2014**, *16*, 1338–1341. [CrossRef]
44. Molander, G.A.; Yun, C.S.; Ribagorda, M.; Biolatto, B. B-alkyl suzuki-miyaura cross-coupling reactions with air-stable potassium alkyltrifluoroborates. *J. Org. Chem.* **2003**, *68*, 5534–5539. [CrossRef]
45. Choi, J.; Fu, G.C. Transition metal-catalyzed alkyl-alkyl bond formation: Another dimension in cross-coupling chemistry. *Science* **2017**, *356*, 1. [CrossRef]

46. Crudden, C.M.; Glasspoole, B.W.; Lata, C.J. Expanding the scope of transformations of organoboron species: Carbon–carbon bond formation with retention of configuration. *Chem. Commun.* **2009**, *44*, 6704–6716. [CrossRef]
47. Ohmura, T.; Awano, T.; Suginome, M. Stereospecific Suzuki-Miyaura coupling of chiral α-(Acylamino) benzylboronic esters with inversion of configuration. *J. Am. Chem. Soc.* **2010**, *132*, 13191–13193. [CrossRef] [PubMed]
48. Buchspies, J.; Szostak, M. Recent advances in acyl suzuki cross-coupling. *Catalysts* **2019**, *9*, 53. [CrossRef]
49. Blangetti, M.; Rosso, H.; Prandi, C.; Deagostino, A.; Venturello, P. Suzuki-miyaura cross-coupling in acylation reactions, scope and recent developments. *Molecules* **2013**, *18*, 1188–1213. [CrossRef]
50. Cheng, H.G.; Chen, H.; Liu, Y.; Zhou, Q. The Liebeskind–Srogl Cross-Coupling Reaction and its Synthetic Applications. *Asian J. Org. Chem.* **2018**, *7*, 490–508. [CrossRef]
51. Takise, R.; Muto, K.; Yamaguchi, J. Cross-coupling of aromatic esters and amides. *Chem. Soc. Rev.* **2017**, *46*, 5864–5888. [CrossRef]
52. Guo, L.; Rueping, M. Transition-Metal-Catalyzed Decarbonylative Coupling Reactions: Concepts, Classifications, and Applications. *Chem. Eur. J.* **2018**, *24*, 7794–7809. [CrossRef]
53. Roy, D.; Uozumi, Y. Recent Advances in Palladium-Catalyzed Cross-Coupling Reactions at ppm to ppb Molar Catalyst Loadings. *Adv. Synth. Catal.* **2018**, *360*, 602–625. [CrossRef]
54. Percec, V.; Bae, J.-Y.; Hill, D.H. Aryl Mesylates in Metal Catalyzed Homocoupling and Cross-Coupling Reactions. 2. Suzuki-Type Nickel-Catalyzed Cross-Coupling of Aryl Arenesulfonates and Aryl Mesylates with Arylboronic Acids. *J. Org. Chem.* **1995**, *60*, 1060–1065. [CrossRef]
55. Han, F.-S. Transition-metal-catalyzed Suzuki–Miyaura cross-coupling reactions: A remarkable advance from palladium to nickel catalysts. *Chem. Soc. Rev.* **2013**, *42*, 5270–5298. [CrossRef]
56. Dong, L.; Wen, J.; Qin, S.; Yang, N.; Yang, H.; Su, Z.; Yu, X.; Hu, C. Iron-Catalyzed Direct Suzuki–Miyaura Reaction: Theoretical and Experimental Studies on the Mechanism and the Regioselectivity. *ACS Catal.* **2012**, *2*, 1829–1837. [CrossRef]
57. Hatakeyama, T.; Hashimoto, T.; Kathriarachchi, K.K.A.D.S.; Zenmyo, T.; Seike, H.; Nakamura, M. Iron-catalyzed alkyl-alkyl Suzuki-Miyaura coupling. *Angew. Chem. Int. Ed.* **2012**, *51*, 8834–8837. [CrossRef] [PubMed]
58. Ansari, R.M.; Bhat, B.R. Schiff base transition metal complexes for Suzuki–Miyaura cross-coupling reaction. *J. Chem. Sci.* **2017**, *129*, 1483–1490. [CrossRef]
59. Miyaura, N.; Yamada, K.; Suzuki, A. A new stereospecific cross-coupling by the palladium-catalyzed reaction of 1-alkenylboranes with 1- alkenyl or 1-alkynyl halides. *Tetrahedron Lett.* **1979**, *20*, 3437–3440. [CrossRef]
60. De Meijere, A.; Brase, S.; Oestreich, M. Metal-Catalyzed Cross-Coupling Reactions and More. In *Metal-Catalyzed Cross-Coupling Reactions and More*; Wiley: New York, NY, USA, 2014.
61. Xu, Z.-Y.; Yu, H.-Z.; Fu, Y. Mechanism of Nickel-Catalyzed Suzuki–Miyaura Coupling of Amides. *Chem. Asian J.* **2017**, *12*, 1765–1772. [CrossRef] [PubMed]
62. Phan, N.T.S.; Van Der Sluys, M.; Jones, C.W. On the nature of the active species in palladium catalyzed Mizoroki-Heck and Suzuki-Miyaura couplings—Homogeneous or heterogeneous catalysis, a critical review. *Adv. Synth. Catal.* **2006**, *348*, 609–679. [CrossRef]
63. Ortuño, M.A.; Lledós, A.; Maseras, F.; Ujaque, G. The transmetalation process in Suzuki-Miyaura reactions: Calculations indicate lower barrier via boronate intermediate. *ChemCatChem* **2014**, *6*, 3132–3138. [CrossRef]
64. Lennox, A.J.J.; Lloyd-Jones, G.C. Transmetalation in the Suzuki-Miyaura coupling: The fork in the trail. *Angew. Chem. Int. Ed.* **2013**, *52*, 7362–7370. [CrossRef]
65. Yunker, L.P.E.; Ahmadi, Z.; Logan, J.R.; Wu, W.; Li, T.; Martindale, A.; Oliver, A.G.; McIndoe, J.S. Real-Time Mass Spectrometric Investigations into the Mechanism of the Suzuki-Miyaura Reaction. *Organometallics* **2018**, *37*, 4297–4308. [CrossRef]
66. Aliprantis, A.O.; Canary, J.W. Observation of Catalytic Intermediates in the Suzuki Reaction by Electrospray Mass Spectrometry. *J. Am. Chem. Soc.* **1994**, *116*, 6985–6986. [CrossRef]
67. Nunes, C.M.; Monteiro, A.L. Pd-Catalyzed Suzuki Cross-Coupling Reaction of Bromostilbene: Insights on the Nature of the Boron Species. *J. Braz. Chem. Soc.* **2007**, *18*, 1443–1447. [CrossRef]
68. Braga, A.A.C.; Ujaque, G.; Maseras, F. A DFT study of the full catalytic cycle of the Suzuki-Miyaura cross-coupling on a model system. *Organometallics* **2006**, *25*, 3647–3658. [CrossRef]

69. Suzuki, A. Cross-coupling reactions via organoboranes. *J. Organomet. Chem.* **2002**, *653*, 83–90. [CrossRef]
70. Chatterjee, A.; Ward, T.R. Recent Advances in the Palladium Catalyzed Suzuki-Miyaura Cross-Coupling Reaction in Water. *Catal. Lett.* **2016**, *146*, 820–840. [CrossRef]
71. Lu, G.P.; Voigtritter, K.R.; Cai, C.; Lipshutz, B.H. Ligand effects on the stereochemical outcome of suzuki-miyaura couplings. *J. Org. Chem.* **2012**, *77*, 3700–3703. [CrossRef] [PubMed]
72. Li, C.; Chen, D.; Tang, W. Addressing the Challenges in Suzuki-Miyaura Cross-Couplings by Ligand Design. *Synlett* **2016**, *27*, 2183–2200.
73. Liu, C.; Li, Y.; Li, Y.; Yang, C.; Wu, H.; Qin, J.; Cao, Y. Efficient Solution-Processed Deep-Blue Organic Light-Emitting Diodes Based on Multibranched Oligofluorenes with a Phosphine Oxide Center. *Chem. Mater.* **2013**, *25*, 3320–3327. [CrossRef]
74. Nishimura, H.; Ishida, N.; Shimazaki, A.; Wakamiya, A.; Saeki, A.; Scott, L.T.; Murata, Y. Hole-Transporting Materials with a Two-Dimensionally Expanded py-System around an Azulene Core for Efficient Perovskite Solar Cells. *J. Am. Chem. Soc.* **2015**, *137*, 15656–15659. [CrossRef]
75. Magano, J.; Dunetz, J.R. Large-scale applications of transition metal-catalyzed couplings for the synthesis of pharmaceuticals. *Chem. Rev.* **2011**, *111*, 2177–2250. [CrossRef]
76. Miyaura, N.; Ishiyama, T.; Ishikawa, M.; Suzuki, A. Palladium-catalyzed cross-coupling reactions of B-alkyl-9-BBN or trialkylboranes with aryl and 1-alkenyl halides. *Tetrahedron Lett.* **1986**, *27*, 6369–6372. [CrossRef]
77. Sato, M.; Miyaura, N.; Suzuki, A. Cross-coupling reaction of alkyl- or arylboronic acid esters with organic halides induced by thallium(I) salts and palladium-catalyst. *Chem. Lett.* **1989**, *8*, 1405–1408. [CrossRef]
78. Saito, B.; Fu, G.C. Alkyl-alkyl Suzuki cross-couplings of unactivated secondary alkyl halides at room temperature. *J. Am. Chem. Soc.* **2007**, *129*, 9602–9603. [CrossRef] [PubMed]
79. Molander, G.A.; Canturk, B. Organotrifluoroborates and monocoordinated palladium complexes as catalysts - A perfect combination for Suzuki-Miyaura coupling. *Angew. Chem. Int. Ed.* **2009**, *48*, 9240–9261. [CrossRef] [PubMed]
80. Walker, S.D.; Barder, T.E.; Martinelli, J.R.; Buchwald, S.L. A rationally designed universal catalyst for Suzuki-Miyaura coupling processes. *Angew. Chem. Int. Ed.* **2004**, *43*, 1871–1876. [CrossRef] [PubMed]
81. Chuang, D.W.; El-Shazly, M.; Chen, C.C.; Chung, Y.M.; D. Barve, B.; Wu, M.J.; Chang, F.R.; Wu, Y.C. Synthesis of 1,5-diphenylpent-3-en-1-yne derivatives utilizing an aqueous B-alkyl Suzuki cross coupling reaction. *Tetrahedron Lett.* **2013**, *54*, 5162–5166. [CrossRef]
82. Yu, D.G.; Li, B.J.; Shi, Z.J. Exploration of new C-O electrophiles in cross-coupling reactions. *Acc. Chem. Res.* **2010**, *43*, 1486–1495. [CrossRef]
83. Tobisu, M.; Chatani, N. Cross-Couplings Using Aryl Ethers via C-O Bond Activation Enabled by Nickel Catalysts. *Acc. Chem. Res.* **2015**, *48*, 1717–1726. [CrossRef]
84. Cornella, J.; Zarate, C.; Martin, R. Metal-catalyzed activation of ethers via C–O bond cleavage: A new strategy for molecular diversity. *Chem. Soc. Rev.* **2014**, *43*, 8081–8097. [CrossRef]
85. Guo, L.; Hsiao, C.C.; Yue, H.; Liu, X.; Rueping, M. Nickel-Catalyzed Csp2-Csp3 Cross-Coupling via C-O Bond Activation. *ACS Catal.* **2016**, *6*, 4438–4442. [CrossRef]
86. Guo, L.; Liu, X.; Baumann, C.; Rueping, M. Nickel-Catalyzed Alkoxy–Alkyl Interconversion with Alkylborane Reagents through C–O Bond Activation of Aryl and Enol Ethers. *Angew. Chem. Int. Ed.* **2016**, *55*, 15415–15419. [CrossRef]
87. Zhang, X.M.; Yang, J.; Zhuang, Q.B.; Tu, Y.Q.; Chen, Z.; Shao, H.; Wang, S.H.; Zhang, F.M. Allylic Arylation of 1,3-Dienes via Hydroboration/Migrative Suzuki-Miyaura Cross-Coupling Reactions. *ACS Catal.* **2018**, *8*, 6094–6099. [CrossRef]
88. Kim, D.E.; Zhu, Y.; Newhouse, T.R. Vinylogous acyl triflates as an entry point to α,β-disubstituted cyclic enones via Suzuki-Miyaura cross-coupling. *Org. Biomol. Chem.* **2019**, *17*, 1796–1799. [CrossRef] [PubMed]
89. Mikagi, A.; Tokairin, D.; Usuki, T. Suzuki-Miyaura cross-coupling reaction of monohalopyridines and L-aspartic acid derivative. *Tetrahedron Lett.* **2018**, *59*, 4602–4605. [CrossRef]
90. Liu, X.; Hsiao, C.C.; Guo, L.; Rueping, M. Cross-Coupling of Amides with Alkylboranes via Nickel-Catalyzed C-N Bond Cleavage. *Org. Lett.* **2018**, *20*, 2976–2979. [CrossRef] [PubMed]

91. Chatupheeraphat, A.; Liao, H.H.; Srimontree, W.; Guo, L.; Minenkov, Y.; Poater, A.; Cavallo, L.; Rueping, M. Ligand-Controlled Chemoselective C(acyl)-O Bond vs C(aryl)-C Bond Activation of Aromatic Esters in Nickel Catalyzed C(sp2)-C(sp3) Cross-Couplings. *J. Am. Chem. Soc.* **2018**, *140*, 3724–3735. [CrossRef] [PubMed]
92. Okuda, Y.; Xu, J.; Ishida, T.; Wang, C.A.; Nishihara, Y. Nickel-Catalyzed Decarbonylative Alkylation of Aroyl Fluorides Assisted by Lewis-Acidic Organoboranes. *ACS Omega* **2018**, *3*, 13129–13140. [CrossRef] [PubMed]
93. Vedejs, E.; Chapman, R.; Fields, S.C.; Lin, S.; Schrimpf, M.R. Conversion of Arylboronic Acids into Potassium Aryltrifluoroborates: Convenient Precursors of Arylboron Difluoride Lewis Acids. *J. Org. Chem.* **1995**, *60*, 3020–3027. [CrossRef]
94. Molander, G.A.; Ito, T. Cross-Coupling Reactions of Potassium Alkyltrifluoroborates with Aryl and 1-Alkenyl Trifluoromethanesulfonates. *Org. Lett.* **2001**, *3*, 393–396. [CrossRef]
95. Darses, S.; Genet, J.P. Potassium organotrifluoroborates: New perspectives in organic synthesis. *Chem. Rev.* **2008**, *108*, 288–325. [CrossRef]
96. Molander, G.A.; Ellis, N. Organotrifluoroborates: Protected Boronic Acids That Expand the Versatility of the Suzuki Coupling Reaction Organotrifluoroborates. *Acc. Chem. Res.* **2007**, *40*, 275–286. [CrossRef]
97. Stefani, H.A.; Cella, R.; Vieira, A.S. Recent advances in organotrifluoroborates chemistry. *Tetrahedron* **2007**, *63*, 3623–3658. [CrossRef]
98. Molander, G.A. Organotrifluoroborates: Another Branch of the Mighty Oak. *J. Org. Chem.* **2015**, *80*, 7837–7848. [CrossRef] [PubMed]
99. Harris, M.R.; Li, Q.; Lian, Y.; Xiao, J.; Londregan, A.T. Construction of 1-Heteroaryl-3-azabicyclo[3.1.0]hexanes by sp3-sp2 Suzuki-Miyaura and Chan-Evans-Lam Coupling Reactions of Tertiary Trifluoroborates. *Org. Lett.* **2017**, *19*, 2450–2453. [CrossRef]
100. Tellis, J.C.; Amani, J.; Molander, G.A. Single-Electron Transmetalation: Photoredox/Nickel Dual Catalytic Cross-Coupling of Secondary Alkyl β-Trifluoroboratoketones and -esters with Aryl Bromides. *Org. Lett.* **2016**, *18*, 2994–2997. [CrossRef] [PubMed]
101. Karimi-Nami, R.; Tellis, J.C.; Molander, G.A. Single-Electron Transmetalation: Protecting-Group-Independent Synthesis of Secondary Benzylic Alcohol Derivatives via Photoredox/Nickel Dual Catalysis. *Org. Lett.* **2016**, *18*, 2572–2575. [CrossRef] [PubMed]
102. Primer, D.N.; Molander, G.A. Enabling the Cross-Coupling of Tertiary Organoboron Nucleophiles through Radical-Mediated Alkyl Transfer. *J. Am. Chem. Soc.* **2017**, *139*, 9847–9850. [CrossRef]
103. Brown, H.C.; Racherla, U.S. Organoboranes. 43. A Convenient, Highly Efficient Synthesis of Triorganylboranes via a Modified Organometallic Route. *J. Org. Chem* **1986**, *51*, 427–432. [CrossRef]
104. Miyaura, N.; Ishiyama, T.; Sasaki, H.; Ishikawa, M.; Sato, M.; Suzuki, A. Palladium-Catalyzed Inter- and Intramolecular Cross-Coupling Reactions of B-Alkyl-9-Borabicyclo[3.3.1]nonane Derivatives with 1-Halo-1-alkenes or Haloarenes. Syntheses of Functionalized Alkenes, Arenes, and Cycloalkenes via a Hydroboration-Coupling Sequence. *J. Am. Chem. Soc.* **1989**, *111*, 314–321.
105. Sun, H.X.; Sun, Z.H.; Wang, B. B-Alkyl Suzuki-Miyaura cross-coupling of tri-n-alkylboranes with arylbromides bearing acidic functions under mild non-aqueous conditions. *Tetrahedron Lett.* **2009**, *50*, 1596–1599. [CrossRef]
106. Wang, B.; Sun, H.X.; Sun, Z.H.; Lin, G.Q. Direct B-alkyl Suzuki-Miyaura cross-coupling of trialkyl-boranes with aryl bromides in the presence of unmasked acidic or basic functions and base-labile protections under mild non-aqueous conditions. *Adv. Synth. Catal.* **2009**, *351*, 415–422. [CrossRef]
107. Monot, J.; Brahmi, M.M.; Ueng, S.; Robert, C.; Murr, M.D.; Curran, D.P.; Malacria, M.; Fensterbank, L.; Lacôte, E. Suzuki—Miyaura Coupling of NHC—Boranes: A New Addition to the C-C Coupling Toolbox. *Org. Lett.* **2009**, *11*, 4914–4917. [CrossRef]
108. Li, H.; Zhong, Y.L.; Chen, C.Y.; Ferraro, A.E.; Wang, D. A Concise and Atom-Economical Suzuki-Miyaura Coupling Reaction Using Unactivated Trialkyl- and Triarylboranes with Aryl Halides. *Org. Lett.* **2015**, *17*, 3616–3619. [CrossRef] [PubMed]
109. Gray, M.; Andrews, I.P.; Hook, D.F.; Kitteringham, J.; Voyle, M. Practical methylation of aryl halides by Suzuki ± Miyaura coupling. *Tetrahedron Lett.* **2000**, *41*, 6237–6240. [CrossRef]
110. Doucet, H. Suzuki-Miyaura cross-coupling reactions of alkylboronic acid derivatives or alkyltrifluoroborates with aryl, alkenyl or alkyl halides and triflates. *Eur. J. Org. Chem.* **2008**, *12*, 2013–2030. [CrossRef]
111. Mu, Y.; Gibbs, R.A. Coupling of isoprenoid triflates with organoboron nucleophiles: Synthesis of all-trans-geranylgeraniol. *Tetrahedron Lett.* **1995**, *36*, 5669–5672. [CrossRef]

112. Zou, G.; Reddy, Y.K.; Falck, J.R. Ag(I)-promoted Suzuki-Miyaura cross-couplings of n-alkylboronic acids. *Tetrahedron Lett.* **2001**, *42*, 7213–7215. [CrossRef]
113. Fall, Y.; Doucet, H.; Santelli, M. Palladium-catalysed Suzuki cross-coupling of primary alkylboronic acids with alkenyl halides. *Appl. Organomet. Chem.* **2008**, *22*, 503–509. [CrossRef]
114. Guo, B.; Fu, C.; Ma, S. Application of LB-Phos·HBF4 in the Suzuki Coupling Reaction of 2-Bromoalken-3-ols with Alkylboronic Acids. *Eur. J. Org. Chem.* **2012**, *2012*, 4034–4041. [CrossRef]
115. Li, C.; Xiao, G.; Zhao, Q.; Liu, H.; Wang, T.; Tang, W. Sterically demanding aryl–alkyl Suzuki–Miyaura coupling. *Org. Chem. Front.* **2014**, *1*, 225–229. [CrossRef]
116. Si, T.; Li, B.; Xiong, W.; Xu, B.; Tang, W. Efficient cross-coupling of aryl/alkenyl triflates with acyclic secondary alkylboronic acids. *Org. Biomol. Chem.* **2017**, *15*, 9903–9909. [CrossRef]
117. Kakiuchi, F.; Usui, M.; Ueno, S.; Chatani, N.; Murai, S. Ruthenium-Catalyzed Functionalization of Aryl Carbon-Oxygen Bonds in Aromatic Ethers with Organoboron Compounds. *J. Am. Chem. Soc.* **2004**, *126*, 2706–2707. [CrossRef]
118. Murai, S.; Kakiuchi, F.; Sekine, S.; Tanaka, Y.; Kamatani, A.; Sonoda, M.; Chatani, N. Efficient catalytic addition of aromatic carbon-hydrogen bonds to olefins. *Nature* **1993**, *366*, 529–531. [CrossRef]
119. Ueno, S.; Mizushima, E.; Chatani, N.; Kakiuchi, F. Direct observation of the oxidative addition of the aryl carbon-oxygen bond to a ruthenium complex and consideration of the relative reactivity between aryl carbon-oxygen and aryl carbon-hydrogen bonds. *J. Am. Chem. Soc.* **2006**, *128*, 16516–16517. [CrossRef] [PubMed]
120. Tobisu, M.; Shimasaki, T.; Chatani, N. Nickel-catalyzed cross-coupling of aryl methyl ethers with aryl boronic esters. *Angew. Chem. Int. Ed.* **2008**, *120*, 4944–4947. [CrossRef]
121. Mancilla, T.; Contreras, R.; Wrackmeyer, B. New bicyclic organylboronic esters derived from iminodiacetic acids. *J. Organomet. Chem.* **1986**, *307*, 1–6. [CrossRef]
122. Li, J.; Grillo, A.S.; Burke, M.D. From Synthesis to Function via Iterative Assembly of N-Methyliminodiacetic Acid Boronate Building Blocks. *Acc. Chem. Res.* **2015**, *48*, 2297–2307. [CrossRef]
123. St. Denis, J.D.; He, Z.; Yudin, A.K. Amphoteric α-Boryl Aldehyde Linchpins in the Synthesis of Heterocycles. *ACS Catal.* **2015**, *5*, 5373–5379. [CrossRef]
124. Duncton, M.A.J.; Singh, R. Synthesis of *trans* -2-(Trifluoromethyl)cyclopropanes via Suzuki Reactions with an *N* -Methyliminodiacetic Acid Boronate. *Org. Lett.* **2013**, *15*, 4284–4287. [CrossRef]
125. Seidel, G.; Fürstner, A. Suzuki reactions of extended scope: The '9-MeO-9-BBN variant' as a complementary format for cross-coupling. *Chem. Commun.* **2012**, *48*, 2055–2070. [CrossRef]
126. Ye, N.; Dai, W.M. An Efficient and Reliable Catalyst System Using Hemilabile Aphos for B-Alkyl Suzuki-Miyaura Cross-Coupling Reaction with Alkenyl Halides. *Eur. J. Org. Chem.* **2013**, *5*, 831–835. [CrossRef]
127. Lee, N.R.; Linstadt, R.T.H.; Gloisten, D.J.; Gallou, F.; Lipshutz, B.H. B-Alkyl sp^3-sp^2 Suzuki-Miyaura Couplings under Mild Aqueous Micellar Conditions. *Org. Lett.* **2018**, *20*, 2902–2905. [CrossRef]
128. Heravi, M.M.; Hashemi, E. Recent applications of the Suzuki reaction in total synthesis. *Tetrahedron* **2012**, *68*, 9145–9178. [CrossRef]
129. Taheri Kal Koshvandi, A.; Heravi, M.M.; Momeni, T. Current Applications of Suzuki-Miyaura Coupling Reaction in The Total Synthesis of Natural Products: An update. *Appl. Organomet. Chem.* **2018**, *32*, e4210. [CrossRef]
130. Wu, Y.D.; Lai, Y.; Dai, W.M. Synthesis of Two Diastereomeric C1–C7 Acid Fragments of Amphidinolactone B Using B-Alkyl Suzuki–Miyaura Cross-Coupling as the Modular Assembly Step. *ChemistrySelect* **2016**, *1*, 1022–1027. [CrossRef]
131. Koch, S.; Schollmeyer, D.; Löwe, H.; Kunz, H. C-Glycosyl amino acids through hydroboration-cross-coupling of exo-glycals and their application in automated solid-phase synthesis. *Chem. Eur. J.* **2013**, *19*, 7020–7041. [CrossRef]
132. Hirai, S.; Utsugi, M.; Iwamoto, M.; Nakada, M. Formal total synthesis of (-)-taxol through Pd-catalyzed eight-membered carbocyclic ring formation. *Chem. Eur. J.* **2015**, *21*, 355–359. [CrossRef]
133. Han, W.; Wu, J. Synthesis of C14–C21 acid fragments of cytochalasin Z $_8$ via anti -selective aldol condensation and *B* -alkyl Suzuki–Miyaura cross-coupling. *RSC Adv.* **2018**, *8*, 3899–3902. [CrossRef]
134. Tungen, J.E.; Aursnes, M.; Hansen, T.V. Stereoselective total synthesis of ieodomycin C. *Tetrahedron* **2014**, *70*, 3793–3797. [CrossRef]

135. Tungen, J.E.; Aursnes, M.; Vik, A. Synthesis of Ieodomycin D. *Synlett* **2016**, *27*, 2497–2499.
136. Xu, G.; Fu, W.; Liu, G.; Senanayake, C.H.; Tang, W. Efficient syntheses of korupensamines A, B and michellamine B by asymmetric Suzuki-Miyaura coupling reactions. *J. Am. Chem. Soc.* **2014**, *136*, 570–573. [CrossRef]
137. Yalcouye, B.; Choppin, S.; Panossian, A.; Leroux, F.R.; Colobert, F. A concise atroposelective formal synthesis of (-)-steganone. *Eur. J. Org. Chem.* **2014**, *2014*, 6285–6294. [CrossRef]

© 2020 by the authors. Licensee MDPI, Basel, Switzerland. This article is an open access article distributed under the terms and conditions of the Creative Commons Attribution (CC BY) license (http://creativecommons.org/licenses/by/4.0/).

Article

Synthesis of a Bcl9 Alpha-Helix Mimetic for Inhibition of PPIs by a Combination of Electrooxidative Phenol Coupling and Pd-Catalyzed Cross Coupling [†]

Martin Vareka [1], Benedikt Dahms [2], Mario Lang [1], Minh Hao Hoang [1,3], Melanie Trobe [1], Hansjörg Weber [1], Maximilian M. Hielscher [2], Siegfried R. Waldvogel [2,*] and Rolf Breinbauer [1,*]

1. Institute of Organic Chemistry, Graz University of Technology, Stremayrgasse 9, A-8010 Graz, Austria; martin.vareka@tugraz.at (M.V.); mario.lang@student.tugraz.at (M.L.); haohm@hcmute.edu.vn (M.H.H.); melanie.trobe@tugraz.at (M.T.); hansjoerg.weber@tugraz.at (H.W.)
2. Department Chemie, Johannes Gutenberg-Universität Mainz, Duesbergweg 10–14, 55128 Mainz, Germany; benedikt.dahms@posteo.de (B.D.); hielscher@uni-mainz.de (M.M.H.)
3. Ho Chi Minh City University of Technology and Education, Vo Van Ngan 01, Linh Chieu Ward, Thu Duc District, Ho Chi Minh City 700000, Vietnam
* Correspondence: waldvogel@uni-mainz.de (S.R.W.); breinbauer@tugraz.at (R.B.)
† Dedicated to Prof. Marko Mihovilovic on the occasion of his 50th birthday.

Received: 12 February 2020; Accepted: 16 March 2020; Published: 19 March 2020

Abstract: Teraryl-based alpha-helix mimetics have resulted in efficient inhibitors of protein-protein interactions (PPIs). Extending the concept to even longer oligoarene systems would allow for the mimicking of even larger interaction sites. We present a highly efficient synthetic modular access to quateraryl alpha-helix mimetics, in which, at first, two phenols undergo electrooxidative dehydrogenative cross-coupling. The resulting 4,4′-biphenol is then activated by conversion to nonaflates, which serve as leaving groups for iterative Pd-catalyzed Suzuki-cross-coupling reactions with suitably substituted pyridine boronic acids. This work, for the first time, demonstrates the synthetic efficiency of using both electroorganic as well as transition-metal catalyzed cross-coupling in the assembly of oligoarene structures.

Keywords: alpha-helix; anode; CH-activation; cross-coupling; electrosynthesis; oligoarene; peptidomimetics; phenol; protein-protein interactions; triflate

---

1. Introduction

Over the last two decades, the inhibition of protein-protein-interactions (PPI) with small molecules has emerged as a challenging but rewarding new paradigm in Chemical Biology and Drug Discovery [1–4]. The challenge is associated with the fact that—in contrast to established drug targets such as enzymes, G protein-coupled receptors (GPCRs), ion channels, etc.—protein-protein-interaction interfaces are characterized by a large, rather flat surface, in which several amino acids distributed over a wide surface area contribute synergistically to the binding of the protein partner. This requires new types of compounds being able to mimic such interaction partners. Among them foldamers [5], stapled peptides [6], and α-helix mimetics [7–13] have turned out to be of particular value. Hamilton and co-workers have demonstrated that trisubstituted linear teraryls can function as α-helix mimetics, displaying the i, i+4 and i+7 amino acid residues in angle and distance characteristic for the α-helix motif within proteins [14]. These teraryl structures have resulted in efficient inhibitors of protein-protein interactions, with the advantages of lower molecular weight, better bioavailability, and hydrolytic stability, when compared with peptide drugs. We could show that such teraryl peptide mimetics can be assembled in a modular way using aryltriflates via Pd-catalyzed cross-coupling reactions [15–18]. In order to address an even larger part of the protein-protein interaction site, we are aiming to synthesize α-helix mimetics in the form of quateraryls featuring four amino acid residues. Ideally, these structures should be accessible from simple starting materials by an iterative cross-coupling process [19]. We envisioned that electrooxidative dehydrogenative coupling of suitably substituted phenols would produce 4,4′-biphenols as building blocks for core fragments [20]. Electroorganic synthesis activates molecules by the simple addition or removal of electrons. Consequently, this method requires no stoichiometric reagents. Currently, this methodology exhibits the lowest environmental impact and is considered as inherently safe [21–24]. Upon conversion of the biphenols into sulfonate esters, these structure motifs could be connected with pyridine boronic acids via Pd-catalyzed cross-coupling reactions.

In this manuscript, we report about the implementation of such a strategy, which enabled us to synthesize a quateraryl fragment, which could function as a mimic of the β-catenin/B-cell CLL/lymphoma 9 protein (Bcl9) interaction site.

2. Results and Discussion

As a test case for our synthetic methodology (see Supplementary Materials), we choose the PPI between β-catenin and Bcl9, which is an important regulatory factor in the development of cancer via the Wnt signaling pathway [25]. The β-catenin/Bcl9 PPI has been well characterized, and the group of Verdine has developed stapled peptides addressing this PPI [26]. By analyzing available structural information from the β-catenin/Bcl9 interface [27], we identified Arg-359, Leu-363, Leu-366, Ile-369, and Leu-373 as relevant amino acids of an α-helix structural element. This led us to propose the following quateraryl structures as target molecules (Figure 1).

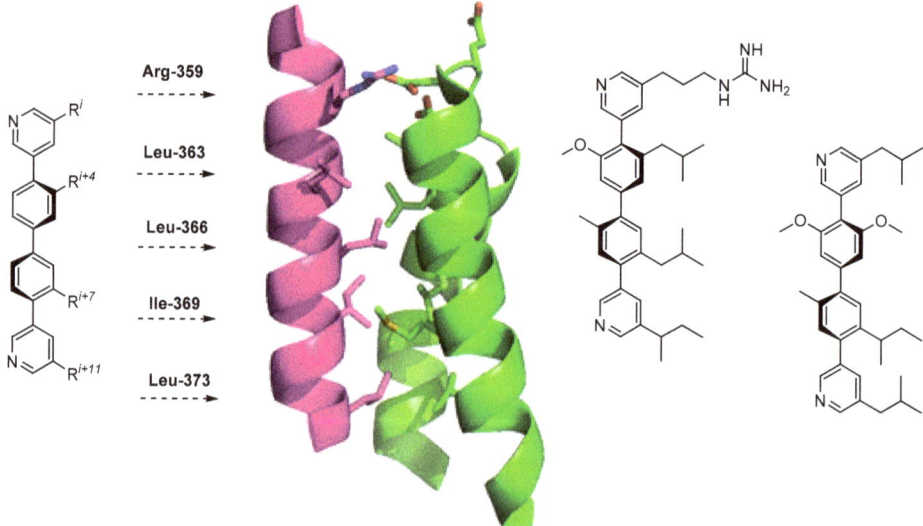

Figure 1. Proposed design for quateraryl mimetics of Bcl9 based on available crystal structure information [27]. The protein structure has been generated using Pymol [28].

2.1. Electrooxidative Cross-Coupling of Phenols

According to our retrosynthetic reasoning, the quateraryls will be assembled from 4,4′-biphenols. In order to have maximum flexibility in the selection of side-chain substituents, the core fragments should be synthesized from differently substituted phenols. In earlier work we could show that symmetric or non-symmetric 4,4′-biphenols can be prepared from suitable ortho-blocked phenols by direct anodic dehydrogenative coupling, using boron-doped diamond (BDD) electrodes and 1,1,1,3,3,3-hexafluoropropan-2-ol (HFIP) as suitable solvent in yields up to 77% [20]. We reasoned that we could improve the selectivity in the electrochemical cross-coupling if we would offer one reaction partner as a free phenol and the second one as a protected phenol [29–31]. For synthetic efficiency, we considered a silyl-protecting group as a good choice. In the event, we tried *tert*-butyldimethylsilyl (TBDMS)-protected phenols **2** and **6** and *tert*-butyldiphenylsilyl (TBDPS)-protected phenol **4** in the coupling with 2,6-dimethoxyphenol (**1**). The yields for the cross-coupling reactions were with 23–27% rather low (Scheme 1, left column). As a comparison, the yields for the unprotected building blocks are displayed (Scheme 1, right column), which, except for **13**, provided the 4,4′-biphenols in yields >60%.

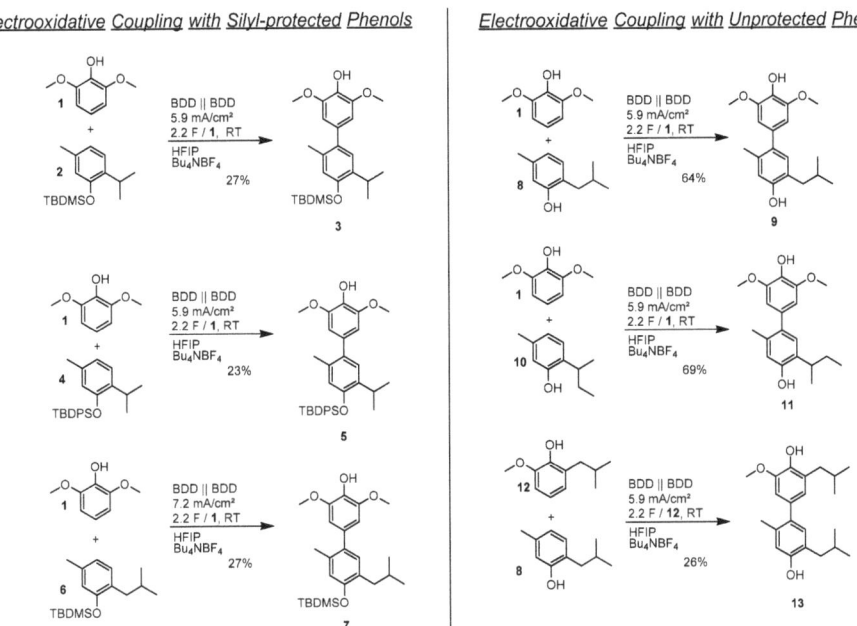

Scheme 1. Electrooxidative cross-coupling of phenols and phenolethers forming 4,4′-biphenol building blocks.

In addition to its poor coupling yields, the silyl-monoprotected building blocks could also not be successfully used in the subsequent nonaflation/Pd-cross coupling steps. Therefore, we preferred to use the unprotected 4,4′-biphenols as core fragments for the assembly of our target quateraryls, which would make the synthetic route even more efficient and shorter.

2.2. Synthesis of Pyridine Boronic Acids

For the final assembly of our target structures we would need pyridine boronic acids featuring the side chain of leucine (**16**), isoleucine (**19**), and arginine (Scheme 2). The leucine pyridine boronic acid was produced in an efficient two-step synthesis starting from 3,5-dichloropyridine (**14**). Fe-catalyzed Kochi-Fürstner cross-coupling [32] of **14** with isobutyl-Grignard delivered **15** in a 53% yield, which could be converted to leucine pyridine boronic acid **16** via Miyaura-borylation with Pd/XPhos in an excellent yield of 93%. For the synthesis of the isoleucine pyridine boronic acid ester **19** a Negishi-coupling strategy was chosen. Starting from 3,5-dibromopyridine (**17**) Negishi-coupling with the in-situ prepared 2-butyl zinc reagent furnished pyridine **18** in a 34% yield. Activation of **18** with a Knochel-Grignard [33] and an electrophilic quench with (pin)BO^{i}Pr resulted in an isoleucine boronic acid ester **19** in a 50% yield. In previous work, we have realized that an arginine building block would be very difficult to handle, not only in the synthesis of building blocks, but also in the assembly of the oligoarenes. Therefore, we preferred to incorporate this building block in a latent alkylnitrile form **23**, which, after oligoarene assembly, can be efficiently converted to the arginine side chain by nitrile reduction, and converting the resulting primary amine with (Boc)$_2$N-guanylation reagent **24**. The Heck reaction of 3,5-dibromopyridine (**17**) with acrylonitrile furnished **20** in a 58% yield. Chemoselective alkene reduction with diimide in situ generated from tosylhydrazide produced **21** in an 86% yield. Building on earlier experience, we chose to convert the bromopyridine **21** to the iodopyridine **22** using the Buchwald–Finkelstein reaction [34] in order to facilitate the planned metalation, with the Knochel–Grignard forming a pyridinyl-Grignard intermediate. Indeed, this transformation and subsequent electrophilic quench with (pin)BO^{i}Pr allowed the isolation of latent Arg-building block (Arg*) **23** in a good yield of 62%.

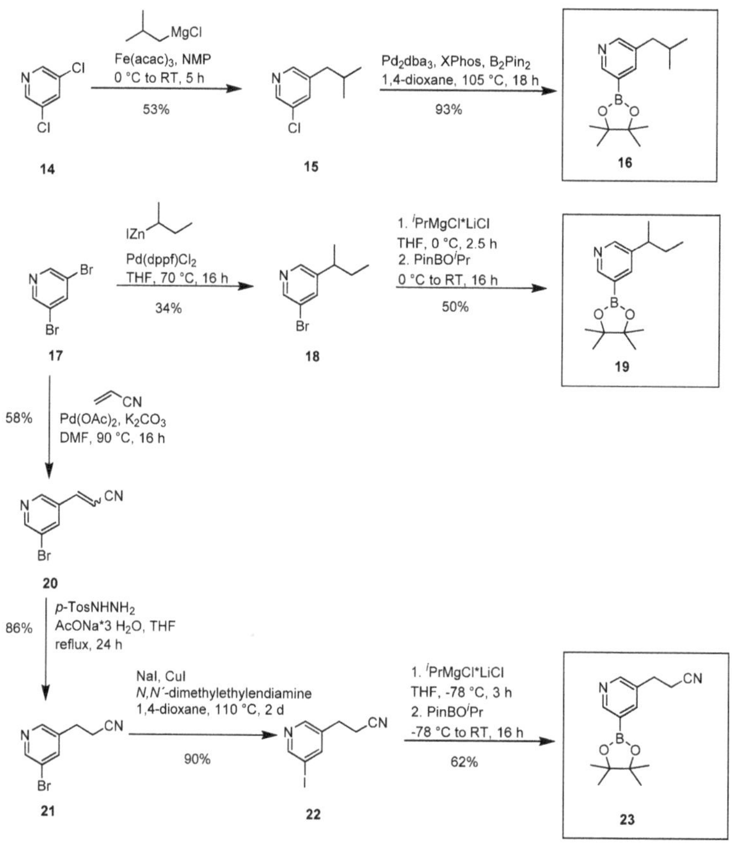

Scheme 2. Synthesis of the pyridine boronic acid ester building blocks.

2.3. Quateraryl Assembly

In order to establish the conditions for the assembly of the quateraryls, we used commercial 3,3′,5,5′-tetramethyl-4,4′-biphenol (**25**) as a model core fragment (Scheme 3). With its two ortho-substituents, it represents a sterically and electronically challenging pattern for subsequent cross-coupling reactions. The methyl substituents are representative of Ala-side chains, giving rise to test compounds, which could serve as control compounds in biological assays following the strategy of an alanine scan widely used in the biochemistry of proteins [35]. Despite considerable effort in optimization, we never succeeded in using the bistriflate of **25** in Pd-catalyzed cross-coupling reactions. We faced considerable side reactions in the form of hydrolysis of the triflate by any type of inorganic base used in the Suzuki-coupling reactions. Therefore, we chose nonaflates as leaving groups, which have been described as a more stable and convenient substrate in Pd-catalyzed cross-coupling reactions [36]. Nonaflation of **25** with nonafluorobutanesulfonylfluoride (NfF) in DCM delivered bisnonaflate **26** in a 62% yield. Suzuki-coupling with 5-methyl-3-pyridine boronic acid ester (**27**) with Pd(dppf)Cl$_2$ as catalyst and K$_2$CO$_3$ as base produced the Ala-Ala-Ala-Ala-quateraryl **28** in a 63% yield. For the synthesis of the asymmetrically substituted Arg-Ala-Ala-Ile quateraryl **30**, Pd(OAc)$_2$/SPhos was chosen as the catalyst. Bisnonaflate **26** was coupled first with Ile-building block **16** and then—after isolation of the teraryl—with the cyanoethyl-building block **23**, using the same catalyst system. As the selectivity of the reaction for the heterocoupling product was only moderate, desired product **29** could only be isolated in a 16% yield. The cyanoethyl group could be converted into the arginine-side chain by first reducing the nitrile to a primary amine with Raney-Ni, followed by reaction with guanylating reagent **24**, producing Arg-Ala-Ala-Ile-quateraryl **30** in a 16% yield over two steps.

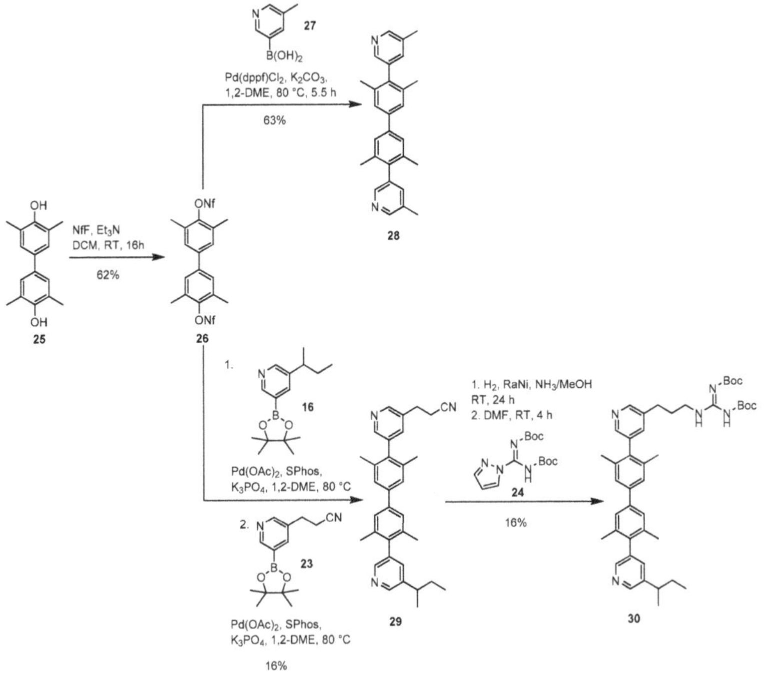

Scheme 3. Synthesis of the quateraryls in the form of Ala-controls.

2.4. Synthesis of Quateraryls as Bcl9-Mimetics

With the productive nonaflate strategy for quateraryl assembly at hand, we could take on the challenge of preparing quateraryls with four different aryl building blocks. As a first target, we selected the Ile-Leu-Leu-Arg*-quateraryl **33** (Scheme 4). Starting from heterocoupling product **13** nonaflation produced **31** in a 19% yield. From the two nonaflate groups in **31**, we expected the nonaflate at the bottom ring for steric and electronic reasons to be more reactive than the one at the top ring, which is flanked by two ortho-substituents, among which one is a strongly electron-donating methoxy group. As expected, the bottom ring nonaflate reacted first in a Suzuki-coupling with Ile-pyridine boronic acid ester **19**, leaving the top ring nonaflate intact for a second Suzuki-coupling with cyanoethyl building block **23**, furnishing target Ile-Leu-Leu-Arg*-quateraryl **33** in decent yields.

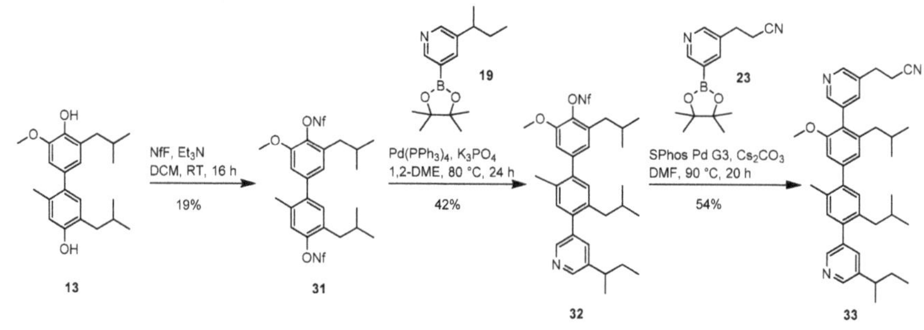

Scheme 4. Synthesis of the Ile-Leu-Leu-Arg*-quateraryl **33**.

Similarly, the Leu-Ile-"MeO"-Leu-quateraryl **36** could be assembled in an impressive 47% overall yield from the bisphenol **11** (Scheme 5). For the coupling of the second nonaflate, again the SPhos Pd G3 catalyst [37] turned out to be very efficient.

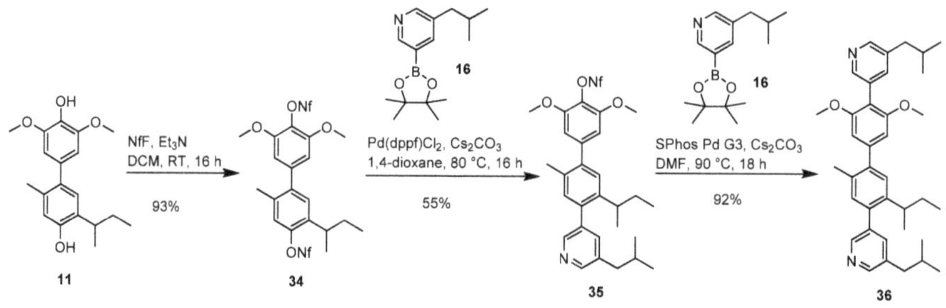

Scheme 5. Synthesis of the Leu-Ile-"MeO"-Leu-quateraryl **36**.

The same precursor also served as the starting material for the synthesis of Ile-Leu-"MeO"-Arg*-quateraryl **39**, which could be synthesized in a 33% overall yield (Scheme 6).

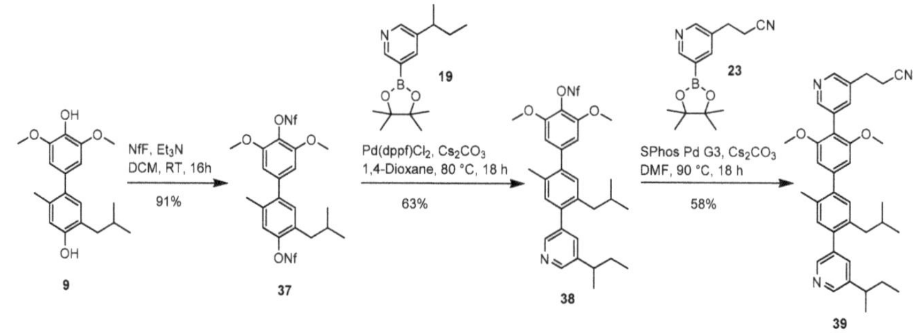

Scheme 6. Synthesis of the Ile-Leu-"MeO"-Arg*-quateraryl **39**.

3. Materials and Methods

Electrochemical Anodic Dehydrogenative Cross-Coupling Reactions

Reaction parameter optimization of anodic cross-coupling reactions was carried out in undivided 5 mL Teflon cells (self-made by the mechanical workshop at JGU Mainz, Germany; or commercially available from IKA, Staufen, Germany as the IKA Screening System), equipped with a Teflon cap for precise alignment (electrode distance: 4.8 mm) of the electrodes. As the electrode material BDD was used (0.3 × 1.0 × 7.0 cm, 15 µm boron-doped diamond layer on silica, commercially available from CONDIAS GmbH, Itzehoe, Germany, DIACHEM^{TM}). Preparative scale electrolysis reactions were carried out in 25 mL undivided beaker-type glass cells with or without cooling jacket (self-made by the mechanical workshop at JGU Mainz), capped with a Teflon plug for precise alignment (electrode distance: 0.8 cm) of the BDD electrodes (0.3 × 2.0 × 6.0 cm, 15 µm boron-doped diamond layer on silica, commercially available from CONDIAS GmbH, Itzehoe, Germany, DIACHEM^{TM}).

4. Conclusions

With the examples shown above, we could for the first time demonstrate the synthetic potential which can be harvested when combining the synthetic efficiency of electrooxidative dehydrogenative cross-coupling of ortho-substituted phenols with the power of Pd-catalyzed cross-coupling reactions. In our research it appeared necessary that the phenols are activated as nonaflates instead of triflates, as the latter showed considerable liabilities in the subsequent Pd-catalyzed reactions due to their hydrolytic lability against bases. In contrast, the nonaflates could be conveniently subjected to Pd-catalyzed cross-coupling reactions. The selectivity could be controlled by electronic and steric effects differentiating the reactivity of the two nonaflate groups. With the synthesis of a Bcl9 quateraryl mimetic, we could highlight this synthetic strategy on a particularly challenging substrate. The overall efficiency was shown in the highly convergent assembly of this quateraryl α-helix mimetic featuring the side chains of Bcl9. We expect that the synthetic methodology reported here will find applications in the synthesis of oligoarene structures, as required in Chemical Biology and Material Sciences.

Supplementary Materials: The following are available online at http://www.mdpi.com/2073-4344/10/3/340/s1, Experimental procedures and full spectroscopic characterization of all synthesized compounds.

Author Contributions: Conceptualization, M.V., S.R.W. and R.B.; development of the electrochemical coupling, B.D., M.V. and S.R.W.; synthesis of building blocks and oligoarene assembly, M.V., M.L., M.T. and M.H.H.; writing—original draft preparation, R.B.; structural assignment via NMR, H.W.; writing—review and editing, M.V., M.M.H. and S.R.W.; funding acquisition, S.R.W. and R.B. All authors have read and agreed to the published version of the manuscript.

Funding: This research was funded by the Austrian Science Fund (FWF) (Project I-2712) to R.B. and by the DFG (Wa1276/14-1) to S.R.W. as a joint D-A-CH project. Support for M.H.H. by the OeAD (Österreichischer Austauschdienst) via an Ernst-Mach-fellowship is gratefully acknowledged.

Acknowledgments: We thank Amber Ford and Sarah Berger for skillful assistance in the lab and for Open Access Funding by the Austrian Science Fund (FWF).

Conflicts of Interest: The authors declare no conflict of interest. The funders had no role in the design of the study; in the collection, analyses, or interpretation of data; in the writing of the manuscript, or in the decision to publish the results".

References

1. Berg, T. Modulation of protein-protein interactions with small organic molecules. *Angew. Chem. Int. Ed.* **2003**, *42*, 2462–2481. [CrossRef]
2. Villoutreix, B.O.; Kuenemann, M.A.; Poyet, J.-L.; Bruzzoni-Giovanelli, H.; Labbé, C.; Lagorce, D.; Sperandio, O.; Miteva, M.A. Drug-Like Protein-Protein Interaction Modulators: Challenges and Opportunities for Drug Discovery and Chemical Biology. *Mol. Inform.* **2014**, *33*, 414–437. [CrossRef] [PubMed]
3. Milroy, L.-G.; Grossmann, T.N.; Hennig, S.; Brunsveld, L.; Ottmann, C. Modulators of protein-protein interactions. *Chem. Rev.* **2014**, *114*, 4695–4748. [CrossRef] [PubMed]

4. Valeur, E.; Guéret, S.M.; Adihou, H.; Gopalakrishnan, R.; Lemurell, M.; Waldmann, H.; Grossmann, T.N.; Plowright, A.T. New Modalities for Challenging Targets in Drug Discovery. *Angew. Chem. Int. Ed.* **2017**, *56*, 10294–10323. [CrossRef] [PubMed]
5. Pelay-Gimeno, M.; Glas, A.; Koch, O.; Grossmann, T.N. Structure-Based Design of Inhibitors of Protein-Protein Interactions: Mimicking Peptide Binding Epitopes. *Angew. Chem. Int. Ed.* **2015**, *54*, 8896–8927. [CrossRef]
6. Walensky, L.D.; Kung, A.L.; Escher, I.; Malia, T.J.; Barbuto, S.; Wright, R.D.; Wagner, G.; Verdine, G.L.; Korsmeyer, S.J. Activation of apoptosis in vivo by a hydrocarbon-stapled BH3 helix. *Science* **2004**, *305*, 1466–1470. [CrossRef]
7. Shaginian, A.; Whitby, L.R.; Hong, S.; Hwang, I.; Farooqi, B.; Searcey, M.; Chen, J.; Vogt, P.K.; Boger, D.L. Design, synthesis, and evaluation of an alpha-helix mimetic library targeting protein-protein interactions. *J. Am. Chem. Soc.* **2009**, *131*, 5564–5572. [CrossRef]
8. Davis, J.M.; Tsou, L.K.; Hamilton, A.D. Synthetic non-peptide mimetics of alpha-helices. *Chem. Soc. Rev.* **2007**, *36*, 326–334. [CrossRef]
9. Cummings, C.G.; Hamilton, A.D. Disrupting protein-protein interactions with non-peptidic, small molecule alpha-helix mimetics. *Curr. Opin. Chem. Biol.* **2010**, *14*, 341–346. [CrossRef]
10. Barnard, A.; Long, K.; Martin, H.L.; Miles, J.A.; Edwards, T.A.; Tomlinson, D.C.; Macdonald, A.; Wilson, A.J. Selective and potent proteomimetic inhibitors of intracellular protein-protein interactions. *Angew. Chem. Int. Ed.* **2015**, *54*, 2960–2965. [CrossRef]
11. Jochim, A.L.; Arora, P.S. Assessment of helical interfaces in protein-protein interactions. *Mol. Biosyst.* **2009**, *5*, 924–926. [CrossRef]
12. Jochim, A.L.; Arora, P.S. Systematic analysis of helical protein interfaces reveals targets for synthetic inhibitors. *ACS Chem. Biol.* **2010**, *5*, 919–923. [CrossRef]
13. Azzarito, V.; Long, K.; Murphy, N.S.; Wilson, A.J. Inhibition of α-helix-mediated protein-protein interactions using designed molecules. *Nat. Chem.* **2013**, *5*, 161–173. [CrossRef] [PubMed]
14. Orner, B.P.; Ernst, J.T.; Hamilton, A.D. Toward proteomimetics: Terphenyl derivatives as structural and functional mimics of extended regions of an alpha-helix. *J. Am. Chem. Soc.* **2001**, *123*, 5382–5383. [CrossRef] [PubMed]
15. Peters, M.; Trobe, M.; Breinbauer, R. A modular synthesis of teraryl-based α-helix mimetics, part 2: Synthesis of 5-pyridine boronic acid pinacol ester building blocks with amino acid side chains in 3-position. *Chemistry* **2013**, *19*, 2450–2456. [CrossRef] [PubMed]
16. Peters, M.; Trobe, M.; Tan, H.; Kleineweischede, R.; Breinbauer, R. A modular synthesis of teraryl-based α-helix mimetics, part 1: Synthesis of core fragments with two electronically differentiated leaving groups. *Chemistry* **2013**, *19*, 2442–2449. [CrossRef] [PubMed]
17. Trobe, M.; Breinbauer, R. Improved and scalable synthesis of building blocks for the modular synthesis of teraryl-based alpha-helix mimetics. *Monatsh Chem* **2016**, *147*, 509–521. [CrossRef]
18. Trobe, M.; Peters, M.; Grimm, S.; Breinbauer, R. The Development of a Modular Synthesis of Teraryl-Based α-Helix Mimetics as Potential Inhibitors of Protein–Protein Interactions. *Synlett* **2014**, *25*, 1202–1214. [CrossRef]
19. Dobrounig, P.; Trobe, M.; Breinbauer, R. Sequential and iterative Pd-catalyzed cross-coupling reactions in organic synthesis. *Monatsh Chem* **2017**, *148*, 3–35. [CrossRef]
20. Dahms, B.; Kohlpaintner, P.J.; Wiebe, A.; Breinbauer, R.; Schollmeyer, D.; Waldvogel, S.R. Selective Formation of 4,4′-Biphenols by Anodic Dehydrogenative Cross- and Homo-Coupling Reaction. *Chemistry* **2019**, *25*, 2713–2716. [CrossRef]
21. Waldvogel, S.R.; Lips, S.; Selt, M.; Riehl, B.; Kampf, C.J. Electrochemical Arylation Reaction. *Chem. Rev.* **2018**, *118*, 6706–6765. [CrossRef] [PubMed]
22. Wiebe, A.; Gieshoff, T.; Möhle, S.; Rodrigo, E.; Zirbes, M.; Waldvogel, S.R. Electrifying Organic Synthesis. *Angew. Chem. Int. Ed.* **2018**, *57*, 5594–5619. [CrossRef] [PubMed]
23. Möhle, S.; Zirbes, M.; Rodrigo, E.; Gieshoff, T.; Wiebe, A.; Waldvogel, S.R. Modern Electrochemical Aspects for the Synthesis of Value-Added Organic Products. *Angew. Chem. Int. Ed.* **2018**, *57*, 6018–6041. [CrossRef] [PubMed]
24. Röckl, J.L.; Pollok, D.; Franke, R.; Waldvogel, S.R. A Decade of Electrochemical Dehydrogenative C,C-Coupling of Aryls. *Acc. Chem. Res.* **2020**, *53*, 45–61. [CrossRef] [PubMed]

25. Hahne, G.; Grossmann, T.N. Direct targeting of β-catenin: Inhibition of protein-protein interactions for the inactivation of Wnt signaling. *Bioorg. Med. Chem.* **2013**, *21*, 4020–4026. [CrossRef]
26. Grossmann, T.N.; Yeh, J.T.-H.; Bowman, B.R.; Chu, Q.; Moellering, R.E.; Verdine, G.L. Inhibition of oncogenic Wnt signaling through direct targeting of β-catenin. *Proc. Natl. Acad. Sci. USA* **2012**, *109*, 17942–17947. [CrossRef]
27. Sampietro, J.; Dahlberg, C.L.; Cho, U.S.; Hinds, T.R.; Kimelman, D.; Xu, W. Crystal structure of a beta-catenin/BCL9/Tcf4 complex. *Mol. Cell* **2006**, *24*, 293–300. [CrossRef]
28. DeLano, W.L. Pymol: An open-source molecular graphics tool. In *CCP4 Newsletter on Protein Crystallography*; 2002; pp. 82–92, Unpublished work.
29. Wiebe, A.; Riehl, B.; Lips, S.; Franke, R.; Waldvogel, S.R. Unexpected high robustness of electrochemical cross-coupling for a broad range of current density. *Sci. Adv.* **2017**, *3*, eaao3920. [CrossRef]
30. Wiebe, A.; Schollmeyer, D.; Dyballa, K.M.; Franke, R.; Waldvogel, S.R. Selective Synthesis of Partially Protected Nonsymmetric Biphenols by Reagent- and Metal-Free Anodic Cross-Coupling Reaction. *Angew. Chem. Int. Ed.* **2016**, *55*, 11801–11805. [CrossRef]
31. Lips, S.; Franke, R.; Waldvogel, S.R. Electrochemical Synthesis of 2-Hydroxy-para-terphenyls by Dehydrogenative Anodic C–C Cross-Coupling Reaction. *Synlett* **2019**, *30*, 1174–1177. [CrossRef]
32. Fürstner, A.; Leitner, A.; Méndez, M.; Krause, H. Iron-catalyzed cross-coupling reactions. *J. Am. Chem. Soc.* **2002**, *124*, 13856–13863. [CrossRef] [PubMed]
33. Baron, O.; Knochel, P. Preparation and selective reactions of mixed bimetallic aromatic and heteroaromatic boron-magnesium reagents. *Angew. Chem. Int. Ed.* **2005**, *44*, 3133–3135. [CrossRef] [PubMed]
34. Klapars, A.; Buchwald, S.L. Copper-catalyzed halogen exchange in aryl halides: An aromatic Finkelstein reaction. *J. Am. Chem. Soc.* **2002**, *124*, 14844–14845. [CrossRef]
35. Weiss, G.A.; Watanabe, C.K.; Zhong, A.; Goddard, A.; Sidhu, S.S. Rapid mapping of protein functional epitopes by combinatorial alanine scanning. *Proc. Natl. Acad. Sci. USA* **2000**, *97*, 8950–8954. [CrossRef]
36. Högermeier, J.; Reissig, H.-U. Nine Times Fluoride can be Good for your Syntheses. Not just Cheaper: Nonafluorobutanesulfonates as Intermediates for Transition Metal-Catalyzed Reactions. *Adv. Synth. Catal.* **2009**, *351*, 2747–2763. [CrossRef]
37. Bruno, N.C.; Tudge, M.T.; Buchwald, S.L. Design and Preparation of New Palladium Precatalysts for C-C and C-N Cross-Coupling Reactions. *Chem. Sci.* **2013**, *4*, 916–920. [CrossRef]

© 2020 by the authors. Licensee MDPI, Basel, Switzerland. This article is an open access article distributed under the terms and conditions of the Creative Commons Attribution (CC BY) license (http://creativecommons.org/licenses/by/4.0/).

Article

Synthesis of Alkynyl Ketones by Sonogashira Cross-Coupling of Acyl Chlorides with Terminal Alkynes Mediated by Palladium Catalysts Deposited over Donor-Functionalized Silica Gel

Miloslav Semler, Filip Horký and Petr Štěpnička *

Department of Inorganic Chemistry, Faculty of Science, Charles University, Hlavova 2030, 128 40 Prague, Czech Republic; majkls@prepere.com (M.S.); filip.blud@seznam.cz (F.H.)
* Correspondence: petr.stepnicka@natur.cuni.cz

Received: 25 September 2020; Accepted: 12 October 2020; Published: 15 October 2020

Abstract: Palladium catalysts deposited over silica gel bearing simple amine ($\equiv$Si(CH$_2$)$_3$NH$_2$) and composite functional amide pendants equipped with various donor groups in the terminal position ($\equiv$Si(CH$_2$)$_3$NHC(O)CH$_2$Y, Y = SMe, NMe$_2$ and PPh$_2$) were prepared and evaluated in Sonogashira-type cross-coupling of acyl chlorides with terminal alkynes to give 1,3-disubstituted prop-2-yn-1-ones. Generally, the catalysts showed good catalytic activity in the reactions of aroyl chlorides with aryl alkynes under relatively mild reaction conditions even without adding a copper co-catalyst. However, their repeated use was compromised by a significant loss of activity after the first catalytic run.

Keywords: deposited catalysts; palladium; functional amides; Sonogashira reaction; alkynyl ketone synthesis

1. Introduction

The first examples of Sonogashira-type cross-coupling of terminal alkynes with acyl chlorides to give alkynyl ketones (Scheme 1) were reported by Crisp and O'Donoghue in 1989 [1], who reacted furoyl chlorides with alkynes in the presence of [PdCl$_2$(PhCN)$_2$]/CuI and triethylamine to produce alkynyl furanyl ketones. With [PdCl$_2$(PPh$_3$)$_2$]/CuI and similar catalysts, this reaction subsequently made it possible to synthesize a number of alkynyl ketones in organic solvents [2,3], in water (when adding sodium dodecyl sulfate as a phase transfer reagent) [4–7] and even in a flow reactor (when using unsupported Pd(OAc)$_2$ as the catalyst) [8].

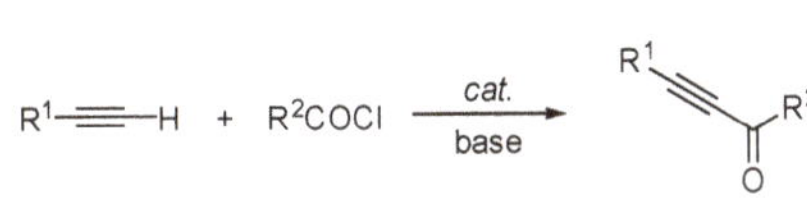

$$R^1{\equiv}H \;+\; R^2COCl \;\xrightarrow[\text{base}]{\text{cat.}}\; \underset{O}{\overset{R^1}{\diagdown}}{\equiv}{\diagup} R^2$$

Scheme 1. Sonogashira cross-coupling of alkynes and acyl chlorides resulting in alkynyl ketones.

Alongside the development of homogenous catalysts, various heterogeneous catalytic systems were devised for this cross-coupling reaction. Wang et al. [9] studied the coupling of aromatic chlorides and cinnamoyl chloride with ethynylbenzene mediated by [PdCl$_2$(PPh$_3$)$_2$]/CuI deposited on KF-alumina under microwave irradiation. Subsequent reports described the use of conventional Pd/C [10], Pd nanoparticles supported by poly(1,4-phenylene sulfide) [11] or by functionalized polystyrene, PS-CH$_2$NHC(S)NHN=C(Ph)C(Me)=N-OH (PS = polystyrene) (without a Cu co-catalyst) [12], and applications of Pd/BaSO$_4$ with a ZnCl$_2$ co-catalyst [13,14] in similar reactions.

In 2009, Tsai et al. [15] reported the application of a Pd-bipyridyl complex grafted onto the mesoporous molecular sieve MCM-41. Coupling reactions of various substrates mediated by this catalyst in neat triethylamine, in the presence of CuI and triphenylphosphine, proceeded satisfactorily at low Pd loading (0.002–0.1 mol.%). More recently, Cai et al. [16] used a related Pd catalyst prepared by depositing Pd(OAc)$_2$ over an MCM-41 surface, modified by $\equiv$Si(CH$_2$)$_3$NHCH$_2$CH$_2$NH$_2$ groups. At 0.2 mol.% Pd loading, and with 0.2 mol.% CuI as a co-catalyst, this material could be reused ten times with only a marginal loss of activity (reaction in triethylamine at 50 °C). Other authors evaluated the related catalysts obtained from supports bearing phosphine-donor groups, e.g., periodic mesoporous silica with $\equiv$CH$_2$CH$_2$PPh$_2$ substituents [17] and polystyrene modified by the –CH$_2$P$^+$Ph$_2$CH$_2$CH$_2$PPh$_2$ Cl$^-$ moieties at the surface [18].

Alkynyl ketones are valuable synthetic building blocks, opening an access to a range of useful compounds, such as intermediates for the synthesis of various heterocycles [19–23], biologically active compounds [24], naturally occurring compounds [25], liquid-crystalline materials [26], and ligands for transition metal ions [27]. In particular, the promising results achieved with deposited catalysts in the cross-coupling of acyl chlorides and alkynes and the wide range of applications of coupling products led us to consider using palladium catalysts deposited over the conventional silica gel bearing donor-substituted amide pendants [28] at the surface (Scheme 2) [29], which were already evaluated in Suzuki-Miyaura biaryl coupling [30]. The results from our study of these catalysts are presented in this contribution, with a particular focus on the reaction scope and a possible influence of the donor moieties within the functional supports that were varied.

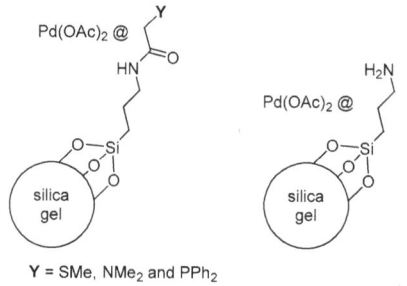

Y = SMe, NMe$_2$ and PPh$_2$

Scheme 2. Deposited catalysts used in this study.

2. Results

2.1. Synthesis of the Catalysts

The deposited catalysts were prepared as reported previously (Scheme 3) [29]. In the first step, freshly calcined, commercial chromatography-grade silica gel (size fraction 63–200 µm) was mixed with (3-aminopropyl)trimethoxysilane in refluxing toluene to afford 3-aminopropylated support **1**. Material **1** was subsequently treated with α-functionalized acetic acids in the presence of peptide coupling agents [31,32], yielding the corresponding amide-functionalized supports **2–4**. In the final step, the resulting materials were treated with palladium(II) acetate in dichloromethane to produce the deposited Pd catalysts **5–7**. As an extension of our previous work, the parent aminopropylated material **1** was also palladated to give material **8** containing only amine functional groups.

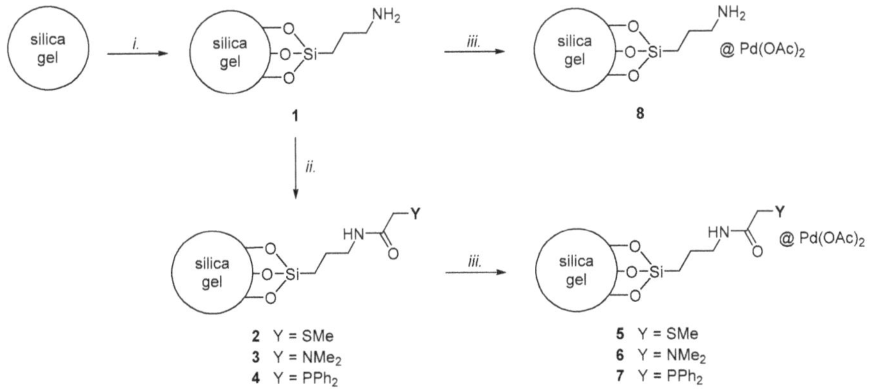

Scheme 3. Preparation of catalysts **5–8**. Legend: *i.* (3-aminopropyl)trimethoxysilane in toluene, refluxing; *ii.* amidation with YCH$_2$CO$_2$H in the presence of peptide coupling agents (1-hydroxybenzotriazole and 1-[3-(dimethylamino)propyl]-3-ethylcarbodiimide (EDC) or the corresponding hydrochloride (EDC·HCl)); *iii.* treatment with Pd(OAc)$_2$ in dichloromethane.

Materials **1–8** were characterized by elemental analysis and infrared (IR) spectroscopy, and the data on **1–7** were compared with those on the previously studied catalysts. While the IR spectra of the newly synthesized materials were virtually identical to those previously reported (see ref. [29]), elemental analysis revealed differences, most likely reflecting the amount of residual adsorbed matter (mostly water). Full characterization data are presented in the Experimental Section.

2.2. Catalytic Assessment

Applications of deposited Pd catalysts to Sonogashira-type coupling of terminal alkynes with acyl chlorides (see Introduction) has been studied considerably less than their use in conventional Sonogashira cross-coupling between alkynes and organic halides [33]. Hence, our initial experiments with catalysts **5–8** aimed to find the optimal reaction conditions for these catalysts and to compare their performance with regard to influence of the varied functional groups modifying the support's surface. As a model reaction, we chose the coupling between equimolar amounts of ethynylbenzene (**9a**) and 4-methylbenzoyl chloride (**10d**), producing 1-(4-methylphenyl)-3-phenyl-2-propyn-1-one (**11ad**, see Scheme 4). The influence of the solvent and base, which are known to strongly affect these reactions (see references in the Introduction), were evaluated first. The screening experiments were performed with 0.5 mol.% of catalyst **5** and 5 mol.% of CuI in neat amines and in mixtures of triethylamine with an organic solvent as well. When using neat morpholine and pyridine, the coupling reaction did not proceed in any appreciable extent. However, when replacing these bases with N-methylmorpholine and N,N-diisopropylethylamine (Figure 1), the yields determined by gas chromatography (GC yields) of the coupling product **11ad** after 8 h at 50 °C were 2% and 10%, respectively. The best (albeit still rather low) yield of 21% after 8 h was achieved in neat triethylamine.

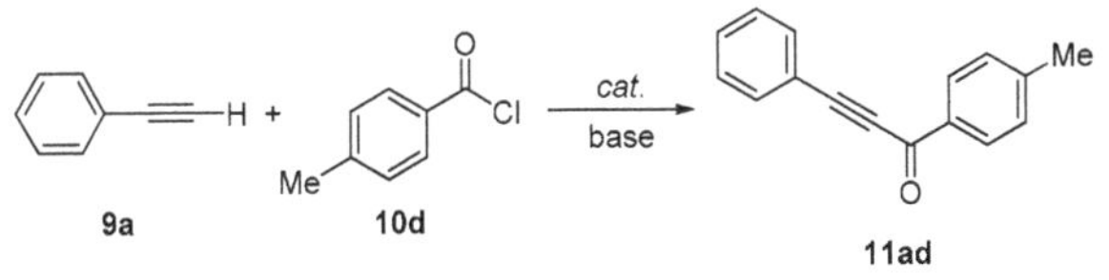

Scheme 4. Coupling reaction used to screen for reaction conditions.

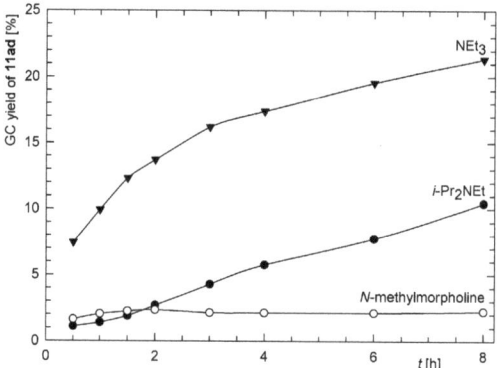

Figure 1. Kinetic profiles for the model coupling reaction performed in neat amines (0.5 mol.% catalyst **5**, 5 mol.% CuI) at 50 °C. Solid lines are added as a visual guide.

Reaction tests performed in organic solvents in the presence of 5 equiv. of triethylamine (Figure 2) revealed a marked acceleration of the coupling reaction in acetonitrile (ca. 60% yield of **11ad** within 3 h at 50 °C). In contrast, reactions in other tested solvents, viz. toluene, 1,4-dioxane, acetone and N,N-dimethylformamide, proceeded less efficiently, achieving lower yields than the aforementioned reaction in neat triethylamine (below 15% after 8 h; Figure 2); no reaction was observed in methanol.

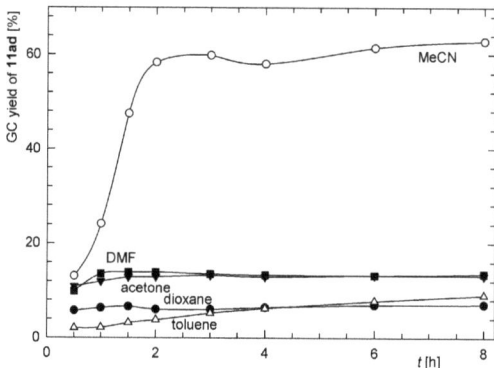

Figure 2. Kinetic profiles for the model coupling reaction performed in organic solvents with added triethylamine (5 equiv. NEt$_3$, 0.5 mol.% catalyst **5**, 5 mol.% CuI) at 50 °C. Legend: MeCN (○), DMF (■), acetone (▼), dioxane (●), toluene (△). The solid lines connecting the experimental points are a visual guide and do not represent any fit of the data.

A subsequent series of experiments was designed to assess the effect of the CuI additive and relative amounts of the starting materials. Rather surprisingly, the reaction performed in neat triethylamine with 0.5 mol.% of catalyst **5** without adding CuI at 50 °C ensued in a higher yield of the coupling product than the similar reaction in the presence of the CuI co-catalyst (5 mol.%; 39% vs. 21%). Consistently, when using acetonitrile as the solvent (with added NEt$_3$, 5 equiv.), the reaction without CuI produced **11ad** in a 78% yield after 8 h, which is a higher yield than that of the reaction performed in the absence of CuI (63%). Subsequently, we determined whether the coupling reaction is affected by the amount of acyl chloride when gradually increasing the amount of 4-toulyl chloride (**10d**) up to 1.5 equiv. As shown in Figure 3, the yield of **11ad** significantly increased with the amount of acyl chloride. With only 1.3 equiv. of **10d**, the GC yields of the coupling product were already virtually quantitative within 1 h of the reaction time.

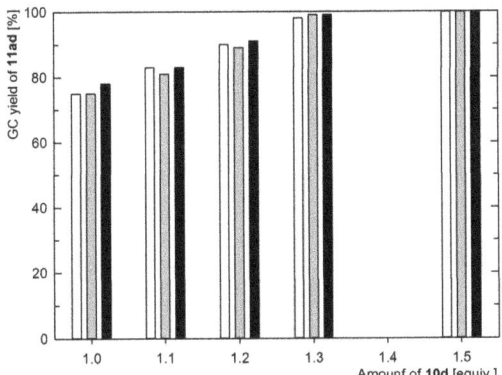

Figure 3. Variation in the gas chromatography (GC) yields of **11ad** observed when changing the amount of acyl chloride in the reaction mixture. Conditions: catalyst **5** (0.5 mol.%), alkyne **9a** (1 equiv.), triethylamine (5 equiv.), dodecane (1 equiv.; internal standard) in acetonitrile solvent at 50 °C. Reaction time: 1 h (white bars), 3 h (grey bars), and 8 h (black bars).

Using 1.5 equiv. of **10d**, we subsequently tried to reduce the catalyst loading. Under these conditions, the reaction proceeded satisfactorily, even in the presence of 0.1 and 0.2 mol.% of the selected model catalyst **5** and at short reaction times, as shown in Figure 4, which compares the GC yields of the coupling product **11ad** achieved over different periods of time. When decreasing the reaction temperature, however, the yield of the coupling product dramatically decreased (100% at 50 °C, 67% at 40 °C and ≈14% at 30 °C after 30 min of the reaction with catalyst **5** and 0.5 mol.% Pd in the reaction mixture).

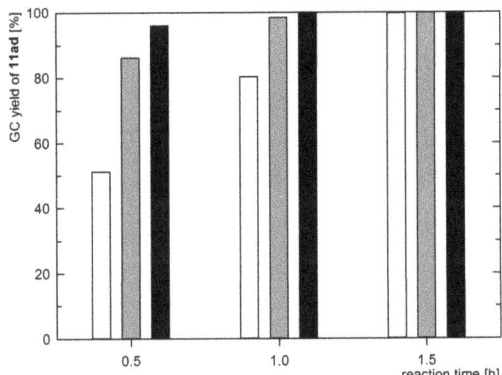

Figure 4. Variation in the GC yields of **11ad** observed upon changing the amount of catalyst **5**. Catalyst loading: 0.1 mol.% (white bars), 0.2 mol.% (grey bars), and 0.5 mol.% (black bars). Conditions: alkyne **9a** (1 equiv.), acyl chloride **10d** (1.5 equiv.), triethylamine (5 equiv.), dodecane (1 equiv.; internal standard) in acetonitrile solvent at 50 °C.

Lastly, we compared all prepared catalysts and palladium(II) acetate under rather harsh reaction conditions (0.1 mol.% Pd, 30 °C reaction temperature). Regrettably, the kinetic profiles presented in Figure 5 clearly indicate that unsupported palladium(II) acetate outperforms all deposited catalysts. Among the deposited catalysts, the lowest efficiency exerted catalyst **7** bearing phosphine groups, whereas the performance of catalysts bearing the S- and N-donor groups (**5** and **6**) was quite similar and slightly better than that of catalyst **8** obtained from the amine-functionalized support.

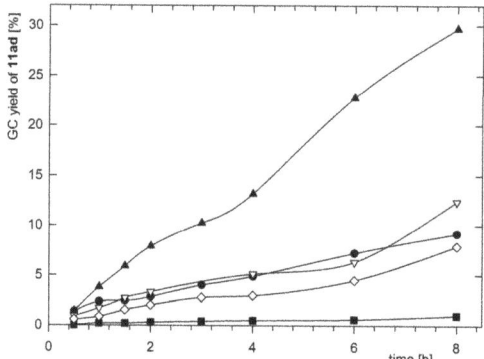

Figure 5. Kinetic profiles for the model coupling reaction performed in the presence of different catalysts: Pd(OAc)$_2$ (▲), catalyst **5** (●), catalyst **6** (▽), catalyst **7** (■), and catalyst **8** (◇). Conditions: 0.1 mol.% Pd, alkyne **9a** (1 equiv.), acyl chloride **10d** (1.5 equiv.), triethylamine (5 equiv.), dodecane (1 equiv.; internal standard) in acetonitrile solvent at 30 °C. The solid lines connecting the experimental points serve as a visual guide and do not represent any fit of the data.

Recycled catalysts **5–8** significantly lost their activity (Figure 6), presumably due to leaching of the deposited metal and to overall catalyst deactivation (the amount of Pd leached out during the first run was only 1%–4% of the initial amount). Notably, CuI (5 mol.%) addition to the reaction mixture increased the stability of the catalysts and even led to an activation of the phosphine-functionalized catalyst **7**, whereas the amount of leached-out Pd remained approximately the same (2–4% during the first run; see the Supporting Information, Table S1). However, the yields of **11ad** obtained with recycled deposited catalysts **5–8**/CuI were still considerably lower than the yields achieved during the first runs and further decreased upon catalyst reuse.

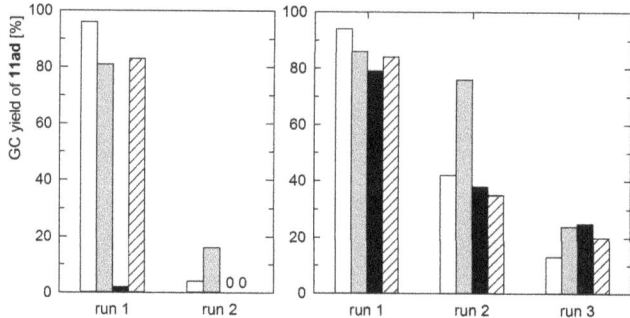

Figure 6. Results of catalytic experiments with fresh and reused catalysts without (left) and with (right) added CuI (5 mol.%): catalyst **5** (white bars), catalyst **6** (grey bars), catalyst **7** (black bars), and catalyst **8** (hatched bars). Conditions: 0.1 mol.% Pd, alkyne **9a** (1 equiv.), acyl chloride **10d** (1.5 equiv.), triethylamine (5 equiv.), dodecane (1 equiv.; internal standard) in acetonitrile at 50 °C for 2 h.

Using catalyst **5** (0.5 mol.% Pd), we also performed reaction scope tests, which are summarized in Table 1. Initially, we focused on the reactions of ethynylbenzene (**9a**) with substituted benzoyl chlorides. In the case of methyl-substituted acyl chlorides, the yields of the coupling products increased with the decrease in steric hindrance. Similar reactions with isomeric nitrobenzoyl chlorides proceeded generally less efficiently and required longer reaction times to achieve isolated yields of the coupling products higher than 50%; the reaction of **9a** with 2-nitrobenzoyl chloride, the most sterically crowded and deactivated acyl chloride, did not proceed. For the acyl chlorides, the substituents with a positive inductive (+*I*) or a mesomeric (+*M*) effect (4-Me, 4-Cl and 4-MeO) apparently facilitated the reaction (isolated yields 85% or higher), whereas the nitro group, with a strong −*M* effect, hampered the cross-coupling. Conversely, the outcome of the coupling reactions between benzoyl chloride (**10a**) and substituted phenylacetylenes (4-Me, 4-MeO and 4-CF$_3$) all proceeded with high isolated yields, in line with the long distance between the substituents in position 4 of the benzene ring and the reaction site, which inevitably minimizes their influence.

Table 1. Summary of the reaction scope tests [a].

Alkyne	Acyl Chloride	Product	Yield (%) [b]
PhC≡CH (**9a**)	2-MeC$_6$H$_4$COCl (**10b**)	**11ab**	66
PhC≡CH (**9a**)	3-MeC$_6$H$_4$COCl (**10c**)	**11ac**	75
PhC≡CH (**9a**)	4-MeC$_6$H$_4$COCl (**10d**)	**11ad**	85
PhC≡CH (**9a**)	2-NO$_2$C$_6$H$_4$COCl (**10e**)	**11ae**	n.d. [d,e]
PhC≡CH (**9a**)	3-NO$_2$C$_6$H$_4$COCl (**10f**)	**11af**	75 [d]
PhC≡CH (**9a**)	4-NO$_2$C$_6$H$_4$COCl (**10g**)	**11ag**	60 [d]
PhC≡CH (**9a**)	4-MeOC$_6$H$_4$COCl (**10h**)	**11ah**	87
PhC≡CH (**9a**)	4-ClC$_6$H$_4$COCl (**10i**)	**11ai**	93
4-MeC$_6$H$_4$C≡CH (**9b**)	PhCOCl (**10a**)	**11ba**	95
4-MeOC$_6$H$_4$C≡CH (**9e**)	PhCOCl (**10a**)	**11ea**	85
4-CF$_3$C$_6$H$_4$C≡CH (**9j**)	PhCOCl (**10a**)	**11ja**	85
PhC≡CH (**9a**)	(*E*)-PhCH=CHCOCl (**10k**)	**11ak**	87
PhC≡CH (**9a**)	PhCH$_2$CH$_2$COCl (**10l**)	**11al**	n.d. [e]
PhC≡CH (**9a**)	*t*-BuCOCl (**10m**)	**11am**	51
PhC≡CH (**9a**)	(2-furanyl)COCl (**10n**)	**11an**	50
PhC≡CH (**9a**)	(2-thienyl)COCl (**10o**)	**11ao**	25 [d]
FcC≡CH (**9m**) [c]	PhCOCl (**10p**)	**11pa**	43 [f]

[a] Conditions: alkyne (1.0 mmol), acyl chloride (1.5 mmol) and triethylamine (5 mmol) were mixed in the presence of catalyst **5** (0.5 mol.% Pd) in acetonitrile (5 mL) at 50 °C for 2 h. [b] Isolated yield after column chromatography. An average of two independent runs is given. [c] Fc = ferrocenyl. [d] Reaction time was extended to 24 h. [e] n.d. = the product was not detected. [f] The reaction was performed with 1.0 mmol of acyl chloride, and the reaction time was extended to 4 h.

The coupling of **9a** with cinnamoyl chloride also proceeded satisfactorily, producing **11ak** in an 87% isolated yield. In contrast, 3-phenylpropanoyl chloride (as a representative of aliphatic acyl chlorides bearing an sp^3 substituent at the acyl group) did not produce any coupling product under analogous conditions. Conversely, pivaloyl chloride was converted into **11am** with an acceptable 51% isolated yield. A similar yield was obtained with 2-furoyl chloride, whereas the reaction with 2-thiophenecarbonyl chloride had a lower yield. The ethynylferrocene/benzoyl chloride pair also displayed a rather sluggish reaction, associated with side processes that were partly suppressed by lowering the amount of the acyl chloride.

In addition to spectroscopic characterization, the structure of **11af** was determined by single-crystal X-ray diffraction analysis. Figure 7 shows the corresponding molecular structure along with selected interatomic distances and angles.

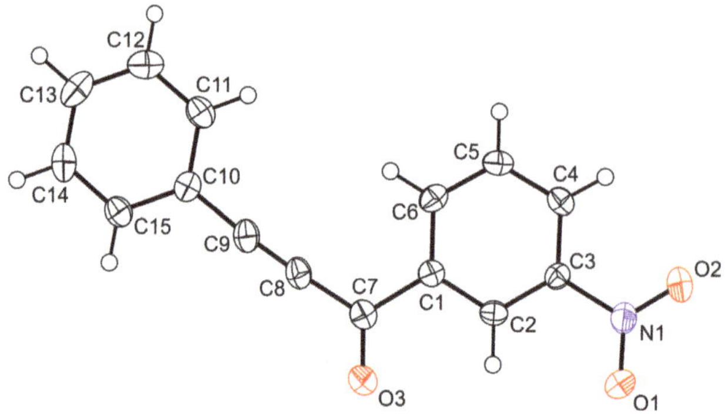

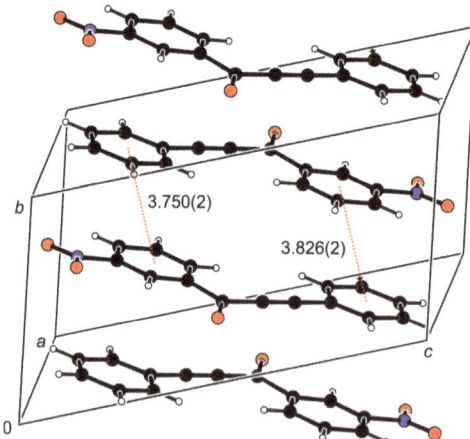

Figure 7. PLATON [34] plot of the molecular structure of **11af** showing the atomic labels and displacement ellipsoids at 50% probability level. Selected distances and angles (in Å and deg): N1=O1 1.224(4), N1=O2 1.229(3), C3-N1 1.468(4), C7=O3 1.223(4), C1-C7 1.492(4), C7-C8 1.447(4), C8-C9 1.205(4), C9-C10 1.433(4); O1=N1=O2 123.4(2), C1-C7-C8 116.7(3), O3=C7-C1/C8 121.6(2)/121.7(2), C7-C8-C9 177.0(3), C8-C9C-10 175.5(3).

The compound crystallizes with the symmetry of the triclinic space group *P*–1 and with one molecule in the asymmetric unit. Parameters describing the molecular geometry of **11af** are unexceptional and in line with the corresponding parameters reported for 1-(4-nitrophenyl)- 3-phenylprop-2-yn-1-one (4-O$_2$NC$_6$H$_4$C(O)C≡CPh) [2,35] and 3-(4-methoxyphenyl)-1-phenylprop-2-yn-1-one (PhC(O)C≡CC$_6$H$_4$OMe-4) [36]. The planes of the benzene rings C(1-6) and C(10-15) in **11af** are essentially coplanar (dihedral angle: 0.4(1)°), and even the nitro group is twisted by only 4.1(3)° with respect to its bonding benzene ring. In the crystal, the individual molecules assemble into columnar stacks of inversion-related molecules (Figure 8) via offset π⋯π stacking interactions of their parallel aromatic rings. These stacks, oriented along the crystallographic *b* axis, are further interconnected in the direction of the crystallographic *a* axis by the C11-H11⋯O3 soft hydrogen bonds (C11⋯O3 = 3.327(3) Å, angle at H11 = 158°).

Figure 8. Section of the columnar stacks in the structure of **11af**. The π⋯π interactions of the parallel benzene rings are indicated by red dotted lines, and the centroid⋯centroid separation is given in Å.

3. Experimental

3.1. Methods and Materials

Infrared spectra were recorded in diffuse reflectance mode using a Fourier-transform infrared spectrometer FTIR Nicolet 6700 (Thermo Fisher Scientific, Waltham, MA, USA; (scan range 400–4000 cm^{-1}, 64 scans, 4 cm^{-1} resolution). The samples analyzed in this study were diluted with KBr (grade for spectroscopy) before the measurement. Nuclear magnetic resonance (NMR) spectra were recorded at 25 °C on a Varian UNITY Inova 400 spectrometer (Palo Alto, CA, USA) operating at 399.95, 100.58 and 376.29 MHz for ^{1}H, ^{13}C and ^{19}F, respectively. Chemical shifts (δ in ppm) are expressed relative to internal tetramethylsilane (^{1}H and ^{13}C) and to external neat CFCl$_3$ (^{19}F). GC analyses were performed with an Agilent 6850 gas chromatograph (Santa Clara, CA, USA) equipped with a DB-5 column (0.18 mm diameter, 50 m length).

Elemental composition of the deposited catalysts was determined using the standard combustion method and a PerkinElmer PE 2400 CHN analyzer (Waltham, MA, USA). The content of palladium in solid samples and in the reaction mixtures was determined by inductively coupled plasma optical emission spectroscopy (ICP-OES) on an IRIS Interpid II instrument (Thermo Electron, Waltham, MA, USA) with axial plasma and ultrasonic CETAC nebulizer U-5000AT+. The samples were dissolved in a mixture of HF with HNO$_3$ (3:2, suprapure from Merck; Kenilworth, NJ, USA) at 50 °C for 15 min and evaporated. The residue was diluted with redistilled water for ^{105}Pd (the wavelength used for the spectrophotometric analysis was 324.270 nm).

Dichloromethane was dried over potassium carbonate and distilled under argon. Other solvents were dried over activated 3 Å molecular sieves. Triethylamine was dried over sodium metal and distilled. Other chemicals were used as obtained from commercial sources (Sigma-Aldrich, St. Louis, MO, USA). Materials **2–7** were prepared as previously described [29]. The analytical data determined for the newly prepared samples are as follows. The IR spectra were identical to those of the authentic samples.

Elemental analysis for **2**: C 6.3, N 1.1, S 0.95 mmol·g^{-1}. Elemental analysis for **3**: C 7.0, N 1.9 mmol·g^{-1}. Elemental analysis for **4**: C 11.3, N 1.1, P 0.77 mmol·g^{-1}. Elemental analysis for **5**: C 7.0, N 0.97, S 0.52, Pd 0.64 mmol·g^{-1}. Elemental analysis for **6**: C 7.6, N 1.7, Pd 0.62 mmol·g^{-1}. Elemental analysis for **7**: C 11.3, N 1.1, P 0.21, Pd 0.43 mmol·g^{-1}.

Catalyst **8** was prepared similarly by direct palladation of material **1**. Thus, palladium(II) acetate (0.449 g, 2.0 mmol) dissolved in dry dichloromethane (10 mL) was added to a suspension of support **1** (2.0 g) [29] in the same solvent (50 mL). After stirring the resulting mixture at room temperature for 1 h, the solid was filtered off and washed with dichloromethane until the washings were colorless. Then, the filter cake was washed a few more times (2-3×) and left to dry in the air.

Characterization data for **8**. IR (DRIFTS): 3648 w, 3243 br w, 1567 m, 1430 w, 1388 w, 1330 vw, 1080 s (Si-O-Si asymetric stretch), 944 vw, 794 m (Si-O-Si symetric stretch), 688 w, 462 (Si-O-Si bending) cm^{-1}. Elemental analysis: C 6.1, N 1.1, Pd 0.58 mmol·g^{-1}.

3.2. Description of the Screening Experiments

A Schlenk tube was successively charged with the catalyst (typically 0.1–0.5 mol.% Pd), CuI (9.5 mg, 5 mol.%; if appropriate), phenylacetylene (102 mg, 1.0 mmol), 4-toluoyl chloride (230 mg, 1.5 mmol) and dodecane (internal standard; 170 mg, 1.0 mmol). The reaction vessel was flushed with nitrogen and sealed. The solvent was introduced (5 mL of pure solvent or 5 mL of a solvent with 697 µL (5 mmol) of triethylamine), and the reaction flask was transferred to a Heidolph Synthesis I parallel reactor pre-heated to the required temperature. Aliquots of the reaction mixture were periodically collected, diluted with saturated aqueous NaHCO$_3$ and centrifuged at 4000 rpm for 5 min. The organic phase was analyzed by gas chromatography.

During recyclation experiments, the reaction mixture obtained after 2 h at 50 °C was diluted with acetone (5 mL) and cooled on ice. A small amount of the liquid phase was separated and used to determine the conversion. The solids were filtered off, washed successively with acetone, methanol (removal of triethylammonium chloride) and acetone again. The filtrate and washings were combined and used to quantify the amount of leached-out metal. The recovered solid was used in the next catalytic experiments.

3.3. Preparative Experiments

A Schlenk tube was charged with the respective alkyne (1.0 mmol) and acyl chloride (1.5 mmol; only 1.0 mmol of the acyl chloride was used in the reaction of ethynylferrocene with benzoyl chloride to avoid decomposition). After flushing the reaction vessel with argon, catalyst **5** (0.5 mol.% Pd) was introduced, followed by dry acetonitrile (5 mL) and anhydrous triethylamine (0.7 mL, ca. 5 mmol). The reaction mixture was stirred at 50 °C for 2 h, diluted with ethyl acetate (10 mL) and cooled on ice. The cold reaction mixture was filtered, and the solid residue was washed with ethyl acetate. The combined organic washings were evaporated under vacuum, leaving a crude reaction product, which was taken up with 1,4-dioxane. Solid NaHCO$_3$ was added ($\approx$0.1 g), and the resulting mixture was stirred at room temperature for 1–7 days to hydrolyze unreacted acyl chloride. The hydrolyzed reaction mixture was evaporated, and the residue was extracted with ethyl acetate. Organic washings were dried over anhydrous MgSO$_4$ and evaporated. Analytically pure coupling products were isolated by column chromatography over silica gel using ethyl acetate-hexane (1:10 or 1:20) as the eluent (dichloromethane was used in the case of **11ag**).

3.4. Analytical Data of the Cross-Coupling Products

1-(2-Tolyl)-3-phenylprop-2-yn-1-one (**11ab**) [37]. ^{1}H NMR (CDCl$_3$): δ 2.68 (s, 3 H, CH$_3$), 7.26–7.29 (m, 1 H, aromatics), 7.33–7.49 (m, 5 H, aromatics), 7.64–7.67 (m, 2 H, aromatics), 8.28–8.32 (m, 1 H, aromatics). ^{13}C{^{1}H} NMR (CDCl$_3$): δ 21.9 (CH$_3$), 88.4 and 91.8 (C≡C), 120.4, 125.9, 128.6, 130.6, 132.2, 132.90, 132.93, 133.2, 135.8, 140.5 (aromatics), 179.8 (C=O).

1-(3-Tolyl)-3-phenylprop-2-yn-1-one (**11ac**) [37]. ^{1}H NMR (CDCl$_3$): δ 2.45 (bq, 3 H, J$_{HH}$ = 0.7 Hz, CH$_3$), 7.41-7.51 (m, 5 H, aromatics), 7.67–7.71 (m, 2 H, aromatics), 8.01–8.06 (m, 2 H, aromatics). ^{13}C{^{1}H} NMR (CDCl$_3$): δ 21.3 (CH$_3$), 87.0 and 92.9 (C≡C), 120.2, 127.1, 128.5, 128.7, 129.8, 130.7, 133.1, 135.0, 136.9 and 138.5 (aromatics), 178.2 (C=O).

1-(4-Tolyl)-3-phenylprop-2-yn-1-one (**11ad**) [37]. ^{1}H NMR (CDCl$_3$): δ 2.45 (s, 3 H, CH$_3$), 7.29–7.33 (m, 2 H, aromatics), 7.40–7.45 (m, 2 H, aromatics), 7.46–7.51 (m, 1 H, aromatics), 7.67–7.71 (m, 2 H, aromatics), 8.10–8.14 (m, 2 H, aromatics). ^{13}C{^{1}H} NMR (CDCl$_3$): δ 21.9 (CH$_3$), 87.0 and 92.6 (C≡C), 120.3, 128.7, 129.4, 129.7, 130.7, 133.0, 134.6 and 145.2 (aromatics), 177.3 (C=O).

1-(3-Nitrophenyl)-3-phenylprop-2-yn-1-one (**11af**) [8]. ^{1}H NMR (CDCl$_3$): δ 7.46 (m, 2 H, aromatics), 7.54 (m, 1 H, aromatics), 7.74 (m, 3 H, aromatics), 8.49 (ddd, J$_{HH}$ = 8.2, 2.3, 1.1 Hz, 1 H, aromatics), 8.53 (dt, J$_{HH}$ = 7.8, 1.4 Hz, 1 H, aromatics), 9.06 (t, J$_{HH}$ = 1.9 Hz, 1 H, aromatics). ^{13}C{^{1}H} NMR (CDCl$_3$): δ 86.2 and 95.3 (C≡C), 119.4, 124.6, 128.2, 128.9, 129.9, 131.5, 133.4, 134.6, 138.1, 148.5 (aromatics), 175.4 (s, C=O). Crystal used for structure determination was grown from chloroform/hexane.

1-(4-Nitrophenyl)-3-phenylprop-2-yn-1-one (**11ag**) [38]. ^{1}H NMR (CDCl$_3$): δ 7.43 and 7.49 (m, 2 H, aromatics), 7.51–7.57 (m, 1 H, aromatics), 7.69–7.74 (m, 2 H, aromatics), 8.38 (m, 4 H, aromatics). ^{13}C{^{1}H} NMR (CDCl$_3$): δ 86.5 and 95.4 (C≡C), 119.4, 123.9, 128.9, 130.5, 131.5, 133.3, 141.0 and 150.9 (aromatics), 175.9 (C=O).

1-(4-Anisyl)-3-phenylprop-2-yn-1-one (**11ah**) [37]. ^{1}H NMR (CDCl$_3$): δ 3.90 (s, 3 H, CH$_3$O), 6.97–7.01 (m, 2 H, aromatics), 7.39–7.50 (m, 3 H, aromatics), 7.66–7.70 (m, 2 H, aromatics), 8.18–8.22 (m, 2 H, aromatics). ^{13}C{^{1}H} NMR (CDCl$_3$): δ 55.6 (CH$_3$), 86.9 and 92.3 (C≡C), 113.9, 120.4, 128.7, 130.3, 130.6, 132.0, 133.0 and 164.5 (aromatics), 176.7 (C=O).

1-(4-Chlorophenyl)-3-phenylprop-2-yn-1-one (**11ai**) [37]. ^{1}H NMR (CDCl$_3$): δ 7.40–7.46 (m, 2 H, aromatics), 7.47-7.52 (m, 3 H, aromatics), 7.67–7.70 (m, 2 H, aromatics), 8.14-8.18 (m, 2 H, aromatics). ^{13}C{^{1}H} NMR (CDCl$_3$): δ 86.6 and 93.6 (C≡C), 119.9, 128.8, 129.0, 130.9, 131.0, 133.1, 135.3 and 140,7 (aromatics), 176.7 (C=O).

3-(4-Tolyl)-1-phenylprop-2-yn-1-one (**11ba**) [38]. ^{1}H NMR (CDCl$_3$): δ 2.41 (s, 3 H, CH$_3$), 7.21–7.25 (m, 2 H, aromatics), 7.49-7.54 (m, 2 H, aromatics), 7.57–7.65 (m, 3 H, aromatics), 8.20–8.24 (m, 2 H, aromatics). ^{13}C{^{1}H} NMR (CDCl$_3$): δ 21.8 (CH$_3$), 86.8 and 93.8 (C≡C), 117.0, 128.6, 129.5, 129.6, 133.1, 134.0, 137.0 and 141,6 (aromatics), 178.1 (C=O).

3-(4-Anisyl)-1-phenylprop-2-yn-1-one (**11ea**) [38]. ^{1}H NMR (CDCl$_3$): δ 3.86 (s, 3 H, CH$_3$O), 6.91–6.96 (m, 2 H, aromatics), 7.49–7.54 (m, 2 H, aromatics), 7.60–7.67 (m, 3 H, aromatics), 8.20–8.24 (m, 2 H, aromatics). ^{13}C{^{1}H} NMR (CDCl$_3$): δ 55.5 (CH$_3$), 86.9 and 94.3 (C≡C), 111.9, 114.4, 128.6, 129.5, 133.9, 135.2, 137.1 and 161.8 (aromatics), 178.1 (C=O).

3-[4-(Trifluoromethyl)phenyl]-1-phenylprop-2-yn-1-one (**11ja**) [39]. ^{1}H NMR (CDCl$_3$): δ 7.51–7.56 (m, 2 H, aromatics), 7.64–7.71 (m, 3 H, aromatics), 7.78–7.81 (m, 2 H, aromatics), 8.20–8.23 (m, 2 H, aromatics). ^{13}C{^{1}H} NMR (CDCl$_3$): δ 88.1 and 90.5 (C≡C), 123.6 (q, $^1J_{FC}$ = 273 Hz, CF$_3$), 124.0, 125.6 (q, $^3J_{FC}$ = 4 Hz), 128.8, 129.6, 132.3 (q, $^2J_{FC}$ = 33 Hz), 133.2, 134.5 and 136.6 (aromatics), 177.7 (C=O) ^{19}F NMR (CDCl$_3$): δ − 63.4 (s).

1-(2-Phenylvinyl)-3-phenylprop-2-yn-1-one (**11ak**) [39]. ^{1}H NMR (CDCl$_3$): δ 6.88 (d, $^3J_{HH}$ = 16.1 Hz, 1 H, CH=), 7.39–7.50 (m, 5 H, aromatics), 7.58–7.68 (m, 4 H, aromatics), 7.91 (d, $^3J_{HH}$ = 16,1 Hz, 1 H, CH=). ^{13}C{^{1}H} NMR (CDCl$_3$): δ 86.6 and 91.5 (C≡C), 120.2, 128.6, 128.7, 129.1, 130.6, 131.2, 133.0, 134.1 and 148.3 (CH=CH and aromatics), 178.2 (C=O).

1-(*t*-Butyl)-3-phenylprop-2-yn-1-one (**11am**) [40]. ^{1}H NMR (CDCl$_3$): δ 1.28 (s, 9 H, CH$_3$), 7.36–7.41 (m, 2 H, aromatics), 7.43–7.48 (m, 1 H, aromatics), 7.56–7.60 (m, 2 H, aromatics). ^{13}C{^{1}H} NMR (CDCl$_3$): δ 26.1 (CH$_3$), 44.9 ((CH$_3$)$_3$C), 86.0 and 92.2 (C≡C), 120.3, 128.6, 130.6 and 133.0 (aromatics), 194.3 (C=O).

1-(2-Furanyl)-3-phenylprop-2-yn-1-one (**11an**) [41]. ^{1}H NMR (CDCl$_3$): δ 6.61 (dd, $^3J_{HH}$ = 3.6 Hz, $^3J_{HH}$ = 1.71 Hz, 1 H, furanyl), 7.39–7.51 (m, 4 H, furanyl and aromatics), 7.63–7.67 (m, 2 H, aromatics), 7.70 (dd, $^3J_{HH}$ = 1.7 Hz, $^4J_{HH}$ = 0.9 Hz, 1 H, furanyl). ^{13}C{^{1}H} NMR (CDCl$_3$): δ 86.2 and 91.0 (C≡C), 112.7, 119.9, 120.9, 128.7, 130.9, 133.1, 148.1, 153.2 (aromatics and furanyl), 164.8 (C=O).

1-(2-Thienyl)-3-phenylprop-2-yn-1-one (**11ao**) [39]. ^{1}H NMR (CDCl$_3$): δ 7.19 (dd, $^3J_{HH}$ = 4.92 Hz, $^3J_{HH}$ = 3.8 Hz, 1 H, thienyl), 7.39–7.51 (m, 3 H, aromatics), 7.65–7.69 (m, 2 H, aromatics), 7.73 (dd, $^3J_{HH}$ = 4.9 Hz, $^4J_{HH}$ = 1.2 Hz, 1 H, thienyl), 8.01 (dd, $^3J_{HH}$ = 3.8 Hz, $^4J_{HH}$ = 1.2 Hz, 1 H, thienyl). ^{13}C{^{1}H} NMR (CDCl$_3$): δ 86.5 and 91.7 (C≡C), 120.0, 128.4, 128.7, 130.9, 133.1, 135.1, 135.3 a 145.0 (thienyl and aromatics), 169.8 (C=O).

3-Ferrocenyl-1-phenylprop-2-yn-1-one (**11pa**) [42]. ^{1}H NMR (CDCl$_3$): δ 4.29 (s, 5 H, C$_5$H$_5$), 4.43 (virtual t, $^3J_{HH}$ = 1.9 Hz, 2 H, C$_5$H$_4$), 4.69 (vt, $^3J_{HH}$ = 1.9 Hz, 2 H, C$_5$H$_4$), 7.49–7.55 (m, 2 H, aromatics), 7.59–7.65 (m, 1 H, aromatics), 8.17–8,21 (m, 2 H, aromatics). ^{13}C{^{1}H} NMR (CDCl$_3$): δ 60.3, 70.5, 70.8 and 73.2 (ferrocene), 85.5 and 96.6 (C≡C), 128.5, 129.4, 133.7, 137.2 (aromatics), 177.6 (C=O).

3.5. Structure Determination

Crystal data for **11af**: C$_{15}$H$_9$NO$_3$, M = 251.23 g·mol^{-1}, light yellow plate, 0.10 × 0.32 × 0.55 mm^3, triclinic, space group p − 1 (no. 2), a = 6.8003(6) Å, b = 7.1934(7) Å, c = 13.471(1) Å; α = 75.075(4)°, β = 79.161(3)°, γ = 69.530(3)°, V = 593.0(1) Å^3, Z = 2, D$_{calc}$ = 1.407 g·mL^{-1}.

Full-set diffraction data were collected with an Apex 2 (Bruker, Billerica, MA, USA) diffractometer equipped with a Cryostream Cooler (Oxford Cryosystems, Oxford, UK) at 150(2) K using graphite-monochromated Mo Kα radiation (λ = 0.71073 Å). The data were corrected for absorption (μ = 0.10 mm^{-1}) using a multi-scan routine incorporated in the diffractometer software. A total of 5295 diffractions was recorded (θ$_{max}$ = 26°, data completeness = 99.3%), of which 2309 were unique (R$_{int}$ = 2.50%) and 1652 were observed according to the I > 2σ(I) criterion.

The structure was solved using direct methods (SHELXS-97 [43]) and refined by a full-matrix least-squares routine based on F^2 (SHELXL-2017 [44]). The non-hydrogen atoms were refined with anisotropic displacement parameters. All hydrogen atoms were included in their theoretical positions and refined as riding atoms with U$_{iso}$(H) assigned to 1.2U$_{eq}$(C). The refinement converged (Δ/σ = 0.000, 172 parameters) to R = 5.77% for the observed, and R = 8.42%, wR = 15.8% for all diffractions. The final difference map revealed no peaks of chemical significance ($\Delta\rho_{max}$ = 0.22, $\Delta\rho_{min}$ = −0.23 e Å^{-3}). CCDC deposition no. 2015269.

4. Conclusions

In summary, we have described the catalytic applications of palladium catalysts deposited over silica gel bearing composite amide-donor functional moieties at the surface in the Sonogashira-type cross-coupling of acyl chlorides with terminal alkynes producing synthetically useful 1,3-disubstituted prop-2-yn-1-ones. The collected data suggest a generally good catalytic performance of these heterogeneous catalysts alone (i.e., without a co-catalyst) in the reactions of aromatic acyl chlorides with aryl alkynes under relatively mild reaction conditions. Nevertheless, a careful optimization is required for achieving good catalytic results, as the catalytic properties are significantly affected by the reaction conditions (solvent and base) and depend on the nature of the functional pendant at the support's surface. Of the tested catalysts, the poorest performance surprisingly exerted catalyst 7 bearing the phosphine moieties, which contrasts with the general notion that phosphine ligands give rise to active cross-coupling catalysts. When recycled, however, the studied catalysts lost their catalytic activity and, therefore, could not be efficiently reused. Very likely, the catalysts serve as a source of catalytically active Pd species that efficiently mediate the cross-coupling reaction but are not redeposited without deactivation.

Supplementary Materials: The following are available online at http://www.mdpi.com/2073-4344/10/10/1186/s1, Table S1: Yields of the coupling product **11ad** and the amount of leached-out Pd in the recycling experiments.

Author Contributions: P.Š. conceived the study and, in collaboration with M.S. and F.H., interpreted the collected data and wrote this article; M.S. and F.H. performed all syntheses and catalytic tests; all authors contributed to the characterization of the coupling products. All authors have read and agreed to the published version of the manuscript.

Funding: This work has been supported by Charles University Research Centre program No. UNCE/SCI/014.

Acknowledgments: The authors thank I. Císařová from the Department of Inorganic Chemistry, Faculty of Science, Charles University for collecting the diffraction data and J. Rohovec from Institute of Geology, Academy of Sciences of the Czech Republic for ICP OES-MS analyses.

Conflicts of Interest: The authors declare no conflict of interest.

References

1. Crisp, G.T.; O'Donoghue, A.I. Palladium-Catalysed Couplings of Halofurans with Activated Alkenes and Terminal Alkynes. A Short Synthesis of Dihydrowyerone. *Synth. Commun.* **1989**, *19*, 1745–1758. [CrossRef]
2. Cox, R.J.; Ritson, D.J.; Dane, T.A.; Berge, J.; Charmant, J.P.H.; Kantacha, A. Room temperature palladium catalysed coupling of acyl chlorides with terminal alkynes. *Chem. Commun.* **2005**, 1037–1039. [CrossRef] [PubMed]
3. Yin, J.; Wang, X.; Liang, Y.; Wu, X.; Chen, B.; Ma, Y. Synthesis of Ferrocenylethynyl Ketones by Coupling of Ferrocenylethyne with Acyl Chlorides. *Synthesis* **2004**, 331–333. [CrossRef]
4. Chen, L.; Li, C.-J. A Remarkably Efficient Coupling of Acid Chlorides with Alkynes in Water. *Org. Lett.* **2004**, *6*, 3151–3153. [CrossRef] [PubMed]
5. Lv, Q.-R.; Meng, X.; Wu, J.-S.; Gao, Y.-J.; Li, C.-L.; Zhu, Q.-Q.; Chen, B.-H. Palladium-, copper- and water solvent facile preparation of ferrocenylethynyl ketones by coupling. *Catal. Commun.* **2008**, *9*, 2127–2130. [CrossRef]
6. Debono, N.; Canac, Y.; Duhayon, C.; Chauvin, R. An Atropo-Stereogenic Diphosphane Ligand with a Proximal Cationic Charge: Specific Catalytic Properties of a Palladium Complex Thereof. *Eur. J. Inorg. Chem.* **2008**, 2991–2999. [CrossRef]
7. Atobe, S.; Masuno, H.; Sonoda, M.; Suzuki, Y.; Shinohara, H.; Shibata, S.; Ogawa, A. Pd-catalyzed coupling reaction of acid chlorides with terminal alkynes using 1-(2-pyridylethynyl)-2-(2-thienylethynyl)benzene ligand. *Tetrahedron Lett.* **2012**, *53*, 1764–1767. [CrossRef]

8. Baxendale, I.R.; Schou, S.C.; Sedelmeier, J.; Ley, S.V. Multi-Step Synthesis by Using Modular Flow Reactors: The Preparation of Yne–Ones and Their Use in Heterocycle Synthesis. *Chem. Eur. J.* **2010**, *16*, 89–94. [CrossRef]
9. Wang, J.-X.; Wei, B.; Huang, D.; Hu, Y.; Bai, L. A Facile Synthesis of Conjugated Acetyl Ketones by Pd(II)-Cu(I) Doped KF/Al$_2$O$_3$-catalyzed Under Microwave Irradiation. *Synth. Commun.* **2001**, *31*, 3337–3343. [CrossRef]
10. Likhar, P.R.; Subhas, M.S.; Roy, M.; Roy, S.; Kantam, M.L. Copper-Free *Sonogashira* Coupling of Acid Chlorides with Terminal Alkynes in the Presence of a Reusable Palladium Catalyst: An Improved Synthesis of 3-Iodochromenones (=3-Iodo-4*H*-1-benzopyran-4-ones). *Helv. Chim. Acta* **2008**, *91*, 259–264. [CrossRef]
11. Santra, S.; Dhara, K.; Ranjan, P.; Bera, P.; Dash, J.; Mandal, S.K. A supported palladium nanocatalyst for copper free acyl Sonogashira reactions: One-pot multicomponent synthesis of N-containing heterocycle. *Green Chem.* **2011**, *13*, 3238–3247. [CrossRef]
12. Bakherad, M.; Keivanloo, A.; Bahramian, B.; Jajarmi, S. Synthesis of Ynones via Recyclable Polystyrene-Supported Palladium(0) Complex Catalyzed Acylation of Terminal Alkynes with Acyl Chlorides under Copper- and Solvent-Free Conditions. *Synlett* **2011**, 311–314. [CrossRef]
13. Yuan, H.; Jin, H.; Li, B.; Shen, Y.; Yue, R.; Shan, L.; Sun, Q.; Zhang, W. Pd/BaSO$_4$-catalyzed cross coupling of acyl chlorides with in situ generated alkynylzinc derivatives for the synthesis of ynones. *Can. J. Chem.* **2013**, *91*, 333–337. [CrossRef]
14. Yuan, H.; Shen, Y.; Yu, S.; Shan, L.; Sun, Q.; Zhang, W. Pd-Catalyzed Cross-Coupling of Acyl Chlorides with In Situ–Generated Alkynylzinc Derivatives for the Synthesis of Ynones. *Synth. Commun.* **2013**, *43*, 2817–2823. [CrossRef]
15. Chen, J.-Y.; Lin, T.-C.; Chen, S.-C.; Chen, A.-J.; Mou, C.-Y.; Tsai, F.-Y. Highly-efficient and recyclable nanosized MCM-41 anchored palladium bipyridyl complex-catalyzed coupling of acyl chlorides and terminal alkynes for the formation of ynones. *Tetrahedron* **2009**, *65*, 10134–10141. [CrossRef]
16. Huang, B.; Yin, L.; Cai, M. A phosphine-free heterogeneous coupling of acyl chlorides with terminal alkynes catalyzed by an MCM-41-immobilized palladium complex. *New J. Chem.* **2013**, *37*, 3137–3144. [CrossRef]
17. Kang, C.; Huang, J.; He, W.; Zhang, F. Periodic mesoporous silica-immobilized palladium(II) complex as an effective and reusable catalyst for water-medium carbon–carbon coupling reactions. *J. Organomet. Chem.* **2010**, *695*, 120–127. [CrossRef]
18. Bakherad, M.; Keivanloo, A.; Bahramian, B.; Rajaie, M. A copper- and solvent-free coupling of acid chlorides with terminal alkynes catalyzed by a polystyrene-supported palladium(0) complex under aerobic conditions. *Tetrahedron Lett.* **2010**, *51*, 33–35. [CrossRef]
19. Sashida, H. An Alternative Facile Preparation of Telluro- and Selenochromones from o-Bromophenyl Ethynyl Ketones. *Synthesis* **1998**, 745–748. [CrossRef]
20. Waldo, J.P.; Larock, R.C. The Synthesis of Highly Substituted Isoxazoles by Electrophilic Cyclization: An Efficient Synthesis of Valdecoxib. *J. Org. Chem.* **2007**, *72*, 9643–9647. [CrossRef]
21. Takahashi, I.; Morita, F.; Kusagaya, S.; Fukaya, H.; Kitagawa, O. Catalytic enantioselective synthesis of atropisomeric 2-aryl-4-quinolinone derivatives with an N–C chiral axis. *Tetrahedron Asymmetry* **2012**, *23*, 1657–1662. [CrossRef]
22. She, Z.; Niu, D.; Chen, L.; Gunawan, M.A.; Shanja, X.; Hersh, W.H.; Chen, Y. Synthesis of Trisubstituted Isoxazoles by Palladium(II)-Catalyzed Cascade Cyclization–Alkenylation of 2-Alkyn-1-one O-Methyl Oximes. *J. Org. Chem.* **2012**, *77*, 3627–3633. [CrossRef] [PubMed]
23. Nordmann, J.; Breuer, N.; Müller, T.J.J. Efficient Consecutive Four-Component Synthesis of 5-Acylpyrid-2-ones Initiated by Copper-Free Alkynylation. *Eur. J. Org. Chem.* **2013**, 4303–4310. [CrossRef]
24. Rao, P.N.P.; Chen, Q.-H.; Knaus, E.E. Synthesis and Structure–Activity Relationship Studies of 1,3-Diarylprop-2-yn-1-ones: Dual Inhibitors of Cyclooxygenases and Lipoxygenases. *J. Med. Chem.* **2006**, *49*, 1668–1683. [PubMed]
25. Leblanc, M.; Fagnou, K. Allocolchicinoid Synthesis via Direct Arylation. *Org. Lett.* **2005**, *7*, 2849–2852. [CrossRef]
26. Zhao, H.-Y.; Guo, L.; Chen, S.-F.; Bian, Z.-X. Synthesis, electrochemistry and liquid crystal properties of 1,2,3-(NH)-triazolylferrocene derivatives. *J. Mol. Struct.* **2013**, *1054–1055*, 164–169. [CrossRef]
27. Specht, Z.G.; Grotjahn, D.B.; Moore, C.E.; Rheingold, A.L. Effects of Hindrance in N-Pyridyl Imidazolylidenes Coordinated to Iridium on Structure and Catalysis. *Organometallics* **2013**, *32*, 6400–6409. [CrossRef]

28. Štěpnička, P. Phosphino-carboxamides: The inconspicuous gems. *Chem. Soc. Rev.* **2012**, *41*, 4273–4305. [CrossRef]
29. Semler, M.; Čejka, J.; Štěpnička, P. Synthesis and catalytic evaluation in the Heck reaction of deposited palladium catalysts immobilized via amide linkers and their molecular analogues. *Catal. Today* **2014**, *227*, 207–214. [CrossRef]
30. Semler, M.; Štěpnička, P. Synthesis of aromatic ketones by Suzuki-Miyaura cross-coupling of acyl chlorides with boronic acids mediated by palladium catalysts deposited over donor-functionalized silica gel. *Catal. Today* **2015**, *243*, 128–133. [CrossRef]
31. El-Faham, A.; Albericio, F. Peptide Coupling Reagents, More than a Letter Soup. *Chem. Rev.* **2011**, *111*, 6557–6602. [CrossRef] [PubMed]
32. Valeur, E.; Bradley, M. Amide bond formation: Beyond the myth of coupling reagents. *Chem. Soc. Rev.* **2009**, *38*, 606–631. [CrossRef] [PubMed]
33. Opanasenko, M.; Štěpnička, P.; Čejka, J. Heterogeneous Pd catalysts supported on silica matrices. *RSC Adv.* **2014**, *4*, 65137–65162. [CrossRef]
34. Spek, A.L. CheckCIF validation ALERTS: What they mean and how to respond. *Acta Crystallogr. Sect. A Crystallogr. Commun.* **2020**, *76*, 1–11. [CrossRef]
35. The Cambridge Crystallographic Data Centre (CCDC). Refcode YANLAB. Available online: https://www.ccdc.cam.ac.uk/ (accessed on 17 July 2020).
36. Smit, J.B.M.; Marais, C.; Malan, F.P.; Bezuidenhout, C.B. Crystal structure of 3-(4-methoxyphenyl)-1-phenylprop-2-yn-1-one, $C_{16}H_{12}O_2$. *Z. Kristallogr. New Cryst. Struct.* **2019**, *234*, 359–360. [CrossRef]
37. Park, A.; Park, K.; Kim, Y.; Lee, S. Pd-Catalyzed Carbonylative Reactions of Aryl Iodides and Alkynyl Carboxylic Acids via Decarboxylative Couplings. *Org. Lett.* **2011**, *13*, 944–947. [CrossRef]
38. Sun, G.; Lei, M.; Hu, L. A facile and efficient method for the synthesis of alkynone by carbonylative Sonogashira coupling using $CHCl_3$ as the CO source. *RCS Adv.* **2016**, *6*, 28442–28446. [CrossRef]
39. Cui, M.; Wu, H.; Jian, J.; Wang, H.; Liu, C.; Daniel, S.; Zeng, Z. Palladium-catalyzed Sonogashira coupling of amides: Access to ynones via C–N bond cleavage. *Chem. Commun.* **2016**, *52*, 12076–12079. [CrossRef]
40. Feng, L.; Hu, T.; Zhang, S.; Xiong, H.-Y.; Zhang, G. Copper-Mediated Deacylative Coupling of Ynones via C–C Bond Activation under Mild Conditions. *Org. Lett.* **2019**, *21*, 9487–9492. [CrossRef]
41. Friel, D.K.; Snapper, M.L.; Hoveyda, A.H. Aluminum-Catalyzed Asymmetric Alkylations of Pyridyl-Substituted Alkynyl Ketones with Dialkylzinc Reagents. *J. Am. Chem. Soc.* **2008**, *130*, 9942–9951. [CrossRef]
42. Tani, T.; Sawatsugawa, Y.; Sano, Y.; Hirataka, Y.; Takahashi, N.; Hashimoto, S.; Sugiura, T.; Tsuchimoto, T. Alkynyl–B(dan)s in Various Palladium-Catalyzed Carbon–Carbon Bond-Forming Reactions Leading to Internal Alkynes, 1,4-Enynes, Ynones, and Multiply Substituted Alkenes. *Adv. Synth. Catal.* **2019**, *361*, 1815–1834. [CrossRef]
43. Sheldrick, G.M. A short history of *SHELX*. *Acta Crystallogr. Sect. A Found. Crystallogr.* **2008**, *64*, 112–122. [CrossRef] [PubMed]
44. Sheldrick, G.M. Crystal structure refinement with *SHELXL*. *Acta Crystallogr. Sect. C Struct. Chem.* **2015**, *71*, 3–8. [CrossRef] [PubMed]

Publisher's Note: MDPI stays neutral with regard to jurisdictional claims in published maps and institutional affiliations.

© 2020 by the authors. Licensee MDPI, Basel, Switzerland. This article is an open access article distributed under the terms and conditions of the Creative Commons Attribution (CC BY) license (http://creativecommons.org/licenses/by/4.0/).

MDPI
St. Alban-Anlage 66
4052 Basel
Switzerland
Tel. +41 61 683 77 34
Fax +41 61 302 89 18
www.mdpi.com

Catalysts Editorial Office
E-mail: catalysts@mdpi.com
www.mdpi.com/journal/catalysts

www.ingramcontent.com/pod-product-compliance
Lightning Source LLC
LaVergne TN
LVHW070545100526
838202LV00012B/390